A
QUANTUM
LIFE

A QUANTUM LIFE

MY UNLIKELY JOURNEY FROM THE STREET TO THE STARS

HAKEEM OLUSEYI

and
JOSHUA HORWITZ

BALLANTINE BOOKS
NEW YORK

To all who reach for the stars,
and to those who teach them.

Contents

I must seek in the stars what was denied to me on earth.

—*Albert Einstein,* in a letter to his secretary and lover, Betty Neumann, in 1924

Note to Readers

As I've looked back over my early life, I've come to recognize that when it comes to personal reminiscence, time and space can bend to conform to the demands of the heart. Nonetheless, after consulting with others who were present at many of the events in this story, I've re-created scenes and dialogue to the best of my ability. I have changed the names of some of the characters to disguise their identities. While re-creating dialogue, I've chosen not to reprise the form of address my friends and I used daily growing up: calling dudes "niggas" and girls "bitches" and "hos." I don't want to normalize self-hate speech for yet another generation of young Black men and women. Otherwise, this memoir is my best effort to truthfully tell the story of my coming of age as a young scientist, and as a young man.

Prologue

SOME YEARS BACK, a magazine profiled my transformation from James Plummer Jr., a nerdy kid from some of America's most deeply scarred urban ghettos, to Hakeem Oluseyi, sole Black physicist inside the Science Mission Directorate at NASA. The article ran with the tagline "The Gangsta Physicist." The handle stuck, and it followed me wherever I went. I understood that it was an eye-catching tag, that it could open doors and young minds that my science degrees alone could not. But over time I grew to resent it. "Gangsta Physicist" didn't describe the totality of who I was, how far I traveled, or how hard I worked to get there.

As a bookish kid, I was an easy target in Watts, Houston's Third Ward, and the Ninth Ward of New Orleans. My gangbanger older cousins taught me the rules of the street by the time I was six: who you could look at in the eyes and who you couldn't, how to tell if the dude walking toward you was Crip or Blood, friend or foe. I developed a sixth sense, what I thought of as my "dark vision," that let me see all the dirt in my hood: where a deal was going down and where the undercover heat was hiding. The scariest time was after sunset, when the predators came out in force.

I was intrigued by the wider universe, including the night sky. But I couldn't see many stars from the streets where I grew up, what with the big-city lights and the smog. And for the sake of my own survival, I didn't want to be caught staring off into space. Celestial navigation wasn't going to help me find my way home without getting beat up or shaken down. By my early teens, I'd adopted a thug persona, walking and talking tough, carrying a gun for protection. But I never joined a gang, and no matter how hard I'd tried to straddle the gangsta-nerd divide, I was still mostly a science geek play-acting a thug.

Looking back at my boyhood in the 1970s, I see a frightened child who lived like a feral animal, surviving day by day on hustle and hope. I had a different nickname in those days. They called me "The Professor" because by the time I was ten years old, I was reading every book I could get my hands on. If anyone had told me I'd grow up to be an actual professor at MIT, UC Berkeley, and the University of Cape Town, I wouldn't have believed them. In my hood, that kind of daydreaming was more likely to get you jacked than help you find your next meal or a safe place to sleep indoors.

By any odds and most calculations, I should not be here writing this book. But here I am. I've learned that no matter how unlikely, anything you imagine for yourself is within the realm of possibility. It's a proven fact of physics. In quantum mechanics, the most improbable of outcomes is called quantum tunneling. If you try to walk through a wall, in all likelihood the wall will win. But there's an infinitesimally small probability that you will find a pathway through the wall. My life has been an oscillating pattern of passing through walls, then walking into the next one and bouncing hard in the opposite direction. I'm living proof that our lives are ruled by the laws of quantum, rather than deterministic, physics.

I don't believe in fate, whether written in the stars or anywhere else. I wasn't destined to find a path through the heaviest hoods in America to an elite career in astrophysics. It could have gone either way for me. At any of a dozen junctures in my life, I could have turned left or right, a gun could have gone off in my hand or in someone else's.

My youth traversed a multiverse of possibilities. In one universe, James Plummer Jr. got shot during a drug deal gone bad and died in the streets of Jackson, Mississippi. In another, nonparallel universe, he found his way to a PhD physics program, learned to design rocket-launched telescopes that could photograph the sun's invisible light spectrum, and became Professor Hakeem Oluseyi.

None of those names—James, Hakeem, Professor, Gangsta Physicist—foretold my journey through the multiverse. But they remind me, like star trails across the night sky, of the limitless possibilities in this quantum life.

—Washington, D.C., 2021

PART ONE
GHETTO CHILD

If I told you that a flower
bloomed in a dark room,
would you trust it?

—Kendrick Lamar,
"Poetic Justice"

1

WAS FOUR YEARS OLD when my family busted apart. What I remember most about that last night together was all the fussing and fighting. When the noise woke us up, my older sister, Bridgette, and I lay in our bed and listened. Bridgette, who was ten, held my hand and tried to soothe me back to sleep. But the shouting just got louder.

I don't know who started the ruckus. Mama and Daddy were always getting into it about this or that, but that night was meaner than usual. It sounded like either Mama had been stepping out on him, like Daddy said, or else that was a filthy lie, like Mama said. By the time Bridgette and I stuck our heads out of our bedroom to look, they'd been hissing and hollering for half an hour.

Just then, Mama picked up a heavy glass ashtray full of butts and threw it at Daddy's head. He ducked and the ashtray hit the wall hard. That's when Daddy punched her. He used to be an amateur boxer, and a pretty good one, according to Aunt Middy. But I'd never seen Daddy take a swing at Mama. That night, he hit her square across the side of the head. She dropped like a sock puppet. As soon as she went down, Daddy kneeled beside her and started crying and apologizing and petting her up, saying sweetheart this and sweetheart that.

But Mama always kept score, and she would always rather get even than make up. Daddy begged her to come to bed, but Mama just turned away from him and shook her head no. Bridgette led me to our bedroom and sang me a lazy-voice lullaby to help me get back to sleep. Mama had other ideas. Later that night, when Daddy was sleeping, she fetched a can of lighter fluid from the barbecue and sprayed it on her side of the bed. When she touched her Zippo to the mattress, Daddy thought he'd woke up in hell, which I guess he had.

When we heard him shrieking, Bridgette and I scrambled out into the hall again, just in time to see Daddy dragging the burning mattress into the backyard. We rushed out behind him through the thick cloud of black smoke that filled the house.

It must have been warm that evening because all the neighbors came out onto their back porches in their underwear to watch. Daddy dumped a pot full of water on the mattress and glared around at the folks on their porches. "What y'all looking at? We got bedbugs is all."

Bridgette led me back through the smoky hallway to our bedroom, shaking her head like she couldn't believe she was living in such a crazy house with such crazy folks. Mama stayed on the back porch with her arms crossed, staring at the smoking mattress and sucking on her Kool cigarette.

The next morning, she told me and Bridgette it was time to pack up and clear out. "Hurry up now, before your Daddy gits home!"

We didn't have any suitcases, so we filled some plastic garbage bags with clothes and whatever else we could grab from out of the house. When we'd pushed everything that would fit into the trunk of our red Ford Maverick, Mama said, "That's enough." I climbed in behind the driver's seat and Bridgette loaded whatever was left into the back seat next to me: a bunch of shoes and bowling trophies, an old blanket and a pile of Mama's dresses still on hangers.

Then we were driving out of New Orleans East and out of the Goose, the only neighborhood I'd ever known. I asked Mama where we were going, and she said, "California." I didn't know what "California" meant. When I asked her if Daddy was coming to California, she just said, "Hush up, now," and lit a Kool. I didn't want to be a crybaby, but my lips started trembling and then my whole head was shaking and snot was running out of my nose. I looked through the back window at the Goose and said goodbye with my eyes.

Bridgette rode shotgun up front, scanning the radio for Motown songs. While Sly and the Family Stone sang "Family Affair," I counted the 185 seconds it took to play. Then I counted the lampposts spinning past as we headed out of town. Counting was always the way I slowed things down when they felt like they were moving too fast. I'd count heartbeats, stairs, or the rotations of a ceiling fan. When we reached the highway I counted the cars driving past us in the other direction. After the sun set I counted the passing headlights till I fell asleep.

I woke after dark and had to pee. Mama pulled over and I climbed out into the chilly night air. There were no cars and no moonlight—just two spouts of headlights pointing forward into the dark. I felt tiny peeing out under the biggest, blackest sky I'd ever seen. Mama was smoking a cigarette alongside the car, and when I asked her why the sky was so big, she told me, "That's a Texas sky. Everything's bigger in Texas." As my eyes adjusted to the dark, the stars overhead grew brighter and brighter, and I felt smaller and smaller.

Then we were rolling west again, and there was nothing left to count along the darkened highway. So I lay out across the pile of Mama's dresses, stared up through the window at a slice of sky, and began to count the stars.

2

MY MAMA—MISS ELAINE to everyone else—wasn't like other mamas. She was what she called "a mover." After we left New Orleans, we didn't stop moving for the rest of that decade. We lived for a while with more than a dozen different relatives on both sides of the family in L.A., Houston, and back in New Orleans. It seemed like Mama was always fussing and fighting with someone. She'd get into it with her boyfriend or her boss or whichever family member we might be crashing with. Then it would be time to load up the Maverick and move on.

It might be an exaggeration to say I grew up in Mama's Maverick. But piling our possessions into that car and driving on to our next apartment—or to whoever's house we'd land at for however long it lasted—was the one predictable thing about those years out west. I went to a different school, or two, every year. I was always the new kid on the block, trying to make friends in a new bad neighborhood. And as soon as I'd made a new friend—usually some quirky sideline kid like me with too much on his mind—it was time to move on. It got to feel normal, letting go of friends without even saying good-bye.

I can mostly recollect where we lived and what grade I was in by the songs that were playing on the radio.

In Watts, when I was in kindergarten, Stevie Wonder's "Superstition" was playing day and night.

By first grade we were living in Houston, when Patti LaBelle was singing "Lady Marmalade."

Second grade was in Pomona, California, with Natalie Cole singing "This Will Be."

Third grade, back in New Orleans, was "Kiss and Say Goodbye" by the Manhattans.

By the time I got to Piney Woods, Mississippi, in fourth grade, I was lip-synching "Strawberry Letter 23" by the Brothers Johnson in front of the bathroom mirror.

It makes me dizzy just to think back at the playlist of places I called home during that tumbleweed time with Mama.

Mama was a mover, but she was also a hard worker. She always had a job. She worked in post offices and in factories. She worked in assembly plants and in clubs. She was a baker in a hospital, ran the cash register at a truck-stop convenience store, and was a security guard. Since she rarely met a morning she could abide, Mama liked to work night shifts. Plus, they were always hiring for the night shift. That's what she said, anyway, when I asked her why she always worked at night.

Mama's jobs changed with the weather, on account of some boss or co-worker who had it in for her and would say something or do something that made her say or do something that made it impossible to go back to work there. "They say they'll mail me my last check," Mama would say as we packed up the Maverick and hit the road again.

IF I HAD TO GUESS, I'd say Mama was always fighting 'cause she was proud, and she couldn't stand it when her pride got bruised. Mama was proud to be Miss Elaine Josephine Alexander, born to a family of Creoles who had never been enslaved in America or anywhere else. Her great-great-grandfather Samuel James Alexander, Jr., was born in Lorraine, France, in 1848 and moved to Santo Domingo before finally settling in the Seventh Ward of New Orleans.

Mama's people weren't formally educated, but they were dignified and hardworking skilled laborers. Her daddy was a plasterer,
like most of the other men in her family. Her mama, Rosemary—
Grandma Rosie to us—was a beautician who wore a white uniform
and ran her own business out of the house.

Mama grew up in a working-class Black neighborhood in New
Orleans East, called the Goose. She was a straight-A student who
liked to read. But she also liked to play the horses. Before Mama
put down a bet, she always tried to get up close to the horse and
whisper-ask in his ear whether or not he was gonna win. That's how
she met Louis Bijou, who groomed horses at the Fairground Race
Track. One day Louis was brushing down a mare that Mama took
an interest in, and the three of them got to whispering and giggling.
Louis talked sugar to Mama but didn't tell her he was married until
three months later—the day she told him she was pregnant. She
broke up with Louis that day—no one got away with lying to
Mama—and she dropped out of school that spring, at age sixteen,
to have her baby.

A year after Bridgette was born, Mama married Wilbur Jones,
better known as Rocky. Mama never liked to cook, and Rocky
cooked a mean gumbo. The marriage only lasted for three months,
though, because Rocky was lazy and didn't want to work. No way
Miss Elaine was gonna stay married to a man who wouldn't work.

Then she met my daddy, James Edward Plummer. One day
Mama was walking home with her best friend, Miss Jeannie, when
James pulled up alongside in his car and offered them a ride. He
sweet-talked both of them, but dropped off Miss Jeannie first so
he could have Mama to himself. After five months of dating, James
told Elaine that he'd bought them a house. A few weeks after
they moved in together, he drove Elaine to the courthouse to get
married—but without telling her first. According to Mama, he was
afraid that she might turn him down if he came right out and proposed. When I was born three years later, in 1967, they named me
after Daddy: James Edward Plummer Jr.

It probably wasn't in the stars for Mama and Daddy. He was country, she was city. She was from a family of proud and free Creoles, while his family came up from slavery and toiled in the soil of Jim Crow Mississippi. He was oil, all soft and sweet-talking. She was vinegar. Things between them were bound to end in flames.

WHEN WE HEADED out west, Mama was in her mid-twenties, a beautiful, brown-skinned young woman with an hourglass figure. She was what was called an "LL," for light-skinned and long-haired, which was a good thing for a woman to be in New Orleans's color-conscious Black community. After a few months in California, she was wearing dashikis and bell-bottoms and her hair was all natural in a big round Angela Davis 'fro. That's how I remember her in those days: lighting candles and incense, smoking joints, and hosting card parties where all sorts of folk would hang around till all hours. Bridgette made me go to bed by midnight, but Mama's parties kept going and going. Bridgette would have to build a wall of pillows around our heads so we could finally get to sleep.

Bridgette was my protector in those days. She helped dress me in the morning, and she put me to bed at night. She cooked me breakfast and dinner and walked me to the corner store or to school. As the youngest and smallest in the family, I was a convenient target for my teenage cousins in L.A. who liked to show off their kung-fu moves to each other. And I was peculiar, always saying weird shit and counting things all the time. "Never mind him," Bridgette would say, stepping in between me and whoever I was annoying. "He's got a vivid imagination, is all."

My preferred posture was to hang upside down on a chair or couch with my feet up in the air—which was ideal for staring into space and daydreaming, my favorite pastimes. And wherever I walked I would look down at my feet and count my steps. Even though I mostly kept my head down, everyone said I had "devil eyes," because they were bright and hazel-colored. I cried more than boys were supposed to cry, and I wet my bed later than most boys.

But most of all, I was a target for bullies because I would rather play games indoors with Bridgette and her girlfriends than practice slap-boxing on the sidewalks with the dudes. The girls would either be playing jacks or jumping rope while singing rhymes or playing four-way hand-clapping games to songs like "Doctor, Doctor" and "Rockin' Robin." My hands were too small for jacks, but I was good at the hand-clapping and jump-roping games.

Papa John—the distant relative we stayed with in Watts when we first got to California—complained that all those girl games were turning me into a sissy. I didn't know what a sissy was, but he said he was gonna beat it out of me. After he'd been drinking his whiskey at night, he'd make me sit on his knee and punch me hard in the chest to try to knock me off. Bridgette would hide his bottle, but he always seemed to find another one.

When I cried to Mama about Papa John hurting me, she said he was just trying to make a man out of me. What kind of man, I wondered, did he want me, a five-year-old boy, to be? A man like Papa John, whose breath always smelled like Crown Royal? A man like my daddy, with his strong arms and sweet, singsong voice?

By the time I turned six, Daddy's voice was all I could remember about him. When the phone rang on holidays or my birthday, I ran to pick it up, hoping I'd hear his voice on the other end. But he never called. And after a while I stopped running after those rings.

3

Mama was working as a nurse's aide in Los Angeles when she met Robert Black. He was born cross-eyed, and Mama took care of him at the hospital after he went there to get his eyes straightened up. She read to him from magazines and talked and joked with him. Mama liked his sweet manner, and she liked that he cooked and worked steady, her two must-be-must-haves in a long-term man.

Robert Black didn't just cook. He actually *was* a cook—in the Merchant Marine. That made him a superstar to us, since we'd been getting by on Bridgette cooking out of cans ever since we'd landed in California. On Saturday afternoons, Robert Black would come over to Papa John's and cook us exotic meals like peppers stuffed with rice and ground beef. I'd wolf them down like a hungry animal, and he'd laugh out loud, his gold front tooth shining out of his handsome, dark-skinned face.

When Robert Black became Mama's high-rotation boyfriend, Bridgette and I started calling him Daddy Robert. Mama always seemed to have a new boyfriend. If he stuck around for more than a few months, Bridgette and I would call him Daddy—Daddy Bob, Daddy Fred, like that, since our real daddys were out of the picture at that point. Then one day Mama up and announced that she and Daddy Robert had got hitched. Bridgette was sore that we hadn't been invited to the wedding, but Mama explained that they'd decided to keep it simple with a justice of the peace at the courthouse, seeing as how they'd both been married a couple of times before. When Bridgette asked her why she got married so often, Mama laughed and said, "Not so often! I only get hitched on leap years."

So it was time to pack up the Maverick and move to Houston, which was Daddy Robert's home port. In Houston, Daddy Robert would come and go, but mostly he was "gone out to sea," which is where it turns out you spend most of your time when you're in the Merchant Marine. .

Mama worked nights in Houston, and for once she didn't have a boss. As soon as we got to town, Daddy Robert bought her a little club called Jackie's Hideaway. I never understood what exactly folks at her club were hiding away from, but Mama explained that Jackie was the name of the woman who used to own the club. The Hideaway had a bar and a small dance floor and a jukebox. On weekends there was a DJ. Bridgette and I were only allowed to visit the Hideaway during the day, and only when Mama took us. We'd sit at the bar and Mama would make us Shirley Temples. If the cook was around, he'd make us a cheeseburger. We never wanted to leave.

Mama worked at the Hideaway most days from midafternoon till after midnight. She didn't get home till three or four in the morning, and then she'd sleep in till noon. Some nights she didn't come home at all. She wasn't a tuck-you-into-bed mama. Or a get-you-up-and-off-to-school mama neither. That was on Bridgette, who shared not only a bedroom but a bed with me most of the time. Bridgette was my real mama in those days, even though she was just a skinny bean pole of a twelve-year-old. I could tell that she didn't like having to be my sister *and* my mama. But she was the one who stuck by me when I needed taking care of.

One night, Daddy Robert was gone out to sea and Mama was going off to work, or out to a card party somewhere. Bridgette told Mama that I was coming down with a fever and that she should stay home to take care of me. Mama just laughed and said, "You know how to take care of Li'l Jame better than I do." Li'l Jame was what she called me, unless she was angry, and then she'd call me JamesPlummerJunior, like it was all one name.

My fever got worse that night. When Bridgette called the Hideaway, they told her that Mama had left early. She didn't come home, and she didn't call. I was burning up so bad the whole bed was soaked with my sweat. Bridgette stayed up with me all night long, fussing over me with a cool wet cloth and running out to the all-night drugstore to buy some Vicks VapoRub to put on my chest when I was having trouble breathing.

At one point when I was deep in my fever dreams, I heard Bridgette praying next to me. "Please, Lord, don't let Li'l Jame die on me tonight. Don't let him die." I opened my eyes to see her kneeling by the bed with her hands clasped together and pressed against her forehead. I tried to reach out to touch her hands, but I was floating above the bed by then. I thought how nice it would be to just float out the bedroom window, and ride Bridgette's prayer up over the rooftops and into the night sky.

My fever finally broke around dawn, which is when Mama glided through the front door, holding her high heels in two fingers slung over her shoulder. Bridgette lit into her like Mama was a teenager and Bridgette was *her* mama. She called her a bad mother who should be ashamed of herself.

"I'm okay, Bridgette," I said, afraid Mama would get mad at her. "I'm okay, Mama."

But Mama just waved her off and laughed. "I knew you'd tend him better than I could." She was right about that.

HOUSTON WAS THE FIRST PLACE I got a peek at how white folks lived. My best friend that year was a white boy named Bobby, who lived down the street. I liked to hang at his house because his family would sit down to eat together at dinnertime, and afterward they would sit around and play card games or watch TV. Nobody shouted or hit anyone. At least, not in front of me.

Bobby's parents taught me how to play bridge, which came easily and naturally to me, even though I was just six. I liked arranging my cards in my hand, lined up in order by suit. The hardest part was holding the cards in a neat fan so nobody could see them. Counting cards and keeping track of them was the easy part—four hands of thirteen cards in four different suits. Where my special powers came into play was seeing what the other players were holding in their hands without actually seeing their cards. First, I watched their eyes as they scanned their cards. Then I watched how they arranged and played their cards. If I held an ace of clubs, I could usually figure out who had the king. If I held five hearts, I could usually figure out where the other eight hearts were hiding. On good days, I felt like the comic-book hero Ultra Boy with his Penetra-Vision, which was even more powerful than Superman's x-ray vision.

Bobby's parents made a big fuss about my excellence at bridge. It was almost as if I was their own kid and they were proud that I could see through the cards. They were happy to school me on all sorts of things. Like eating vegetables. Since Daddy Robert was gone out to sea most of the time, we were eating out of cans and boxes again. But Bobby's family used to eat raw vegetables cut up into bite-sized pieces that you dipped in blue cheese dressing. And cheese and sausage that you sliced and ate on crackers. They ate everything on plates, even snacks between meals. They taught me how to set a table with napkins and how to arrange books on a bookshelf by the alphabet.

I soaked up all the white-folk stuff like it was a secret language from a parallel, but alien, universe that I wasn't supposed to speak anywhere else.

4

DISCOVERING MY SPECIAL POWER at reading and counting cards woke me up to the possibility that I could figure out almost anything. I was intrigued by the invisible and mysterious forces at work in the world around me and inside of things, like the small appliances that Daddy Robert brought back from duty-free ports of call. The toaster, the blender, and the lava lamp were magical objects that drew their power through cords plugged into the walls. Naturally, I wanted to see how the magic happened inside their shiny bodies.

After school, I'd find the pliers and screwdrivers inside Daddy Robert's toolbox and set to work taking apart an appliance, carefully spreading the screws and pieces in a row across the floor. Sometimes they wouldn't fit back together just right. If Mama came in and found me surrounded by leftover pieces, she would fly into a fit and kick the parts across the floor. She'd grab ahold of a belt and give me a whipping. If a belt wasn't handy, she'd use an extension cord, which was a worse kind of whipping. Every whipping—like the one she gave me after she came home to find me taking apart her makeup mirror with the fluorescent lights around the edge—came with a scolding she delivered in time with the whipping: "Why is you" (whip) "always" (whip) "breaking shit" (whip)?!

I was afraid of the whippings, but I couldn't help myself. The gadgets around the house were like a plate of cookies someone sets out and tells you not to touch. I couldn't resist the secrets hidden inside them. How did the TV make pictures? How did the clock radio make music? I especially loved anything with small motors or switches, and anything with tubes inside. They remained mysterious to me—especially electronics with their tiny resistors and capacitors—but their secrets kept calling to me.

The other thing that reliably got me a whipping were my experiments. I became obsessed with how things changed when you heated them to burning. Watching a Pop-Tart turn brown in the toaster oven was more fun than eating it. We had a little electric space heater in the bathroom wall right in front of the toilet, and I loved to turn it on when I was making number two. At night I'd switch off the lights and watch the metal coils glow orange and then red in the dark. One day I wondered what would happen if I rolled up a little ball of my poop and pressed it in there next to the coils. The only way to find out was to experiment.

Without much fuss I was able to make an acorn-sized ball of poop and lodge it against the coils. Then I sat down on the floor and stared at it closely. At first nothing much happened. Then it sizzled and smoked and the surface of the poop ball started changing. I was so mesmerized by what I was seeing that I didn't notice the smell. But then I did. It was awful! I opened the bathroom window, but by then the poop was smoking heavily. So then I opened the door to air out the room. That was a terrible idea. The burnt-poop stink traveled instantly through the whole house. I heard Mama and her friends yell from the front room with disgust, "Good *Lord*! What's that stink?"

Next thing I know Mama and two men were at the bathroom door. They looked at me standing there with my pants half down, and then at the smoldering poop ball in the heater. "Boy! Is you crazy?" one of the men yelled, screwing up his face and unplugging the heater. Mama was good and mad. She snatched me by my shirt, dragged me down the hall to her room, pulled out a belt, and whipped me good.

My next experiment was to light one of Mama's incense sticks and press the glowing end against the plastic shower curtain to see what would happen. To my surprise it burned straight through with almost no resistance, leaving a perfectly round little hole with a burnt inner rim. I had to do more experiments to figure out exactly how long it took an incense cone to burn a wider hole in the curtain. Pretty soon I'd made a complicated Swiss-cheese pattern across half the shower curtain.

Mama went crazy when she saw it. She didn't ask who'd done it. She just came straight after me with her belt held high. I raced ahead of her from room to room, and finally scurried back into the bathroom and latched it with the hook-and-eyelet lock. Then I knelt on the rug in front of the shower and started to pray, just like the nuns had taught me to do in my Sunday school catechism, eyes squeezed shut, hands clasped over my heart. *Please, God, don't let Mama whip me!*

Mama rattled the door and shouted at me to let her in. "I'm gonna break that door down, I swear I am." Then I saw a butter knife sliding through the crack between the door and the door jamb while I pressed down hard on the hook. Mama finally popped the hook and the door swung open. There was Mama standing like a ninja warrior with her belt in her hand. I pulled the bathroom rug over on top of me—but it was no use. Mama whipped me so hard that her leather belt broke in half.

Lesson learned: Try to deflect a whipping with prayer and you'll get a belt broken on your ass.

MOST OF THE TIME, I understood why Mama whipped me. Either for taking things apart or burning them up. Or the time I stole two quarters from her purse and lied to her about it. But sometimes it seemed like she just got angry and needed to hit something. Once, after she got off a phone call with her sisters, I saw her punch a hole in a wall. For a month—until Daddy Robert came home and patched it—that hole just stared out at me like a one-eyed warning sign: Don't mess with Mama.

One day Bridgette and I came home from school to find Mama gone and Daddy Robert in the kitchen cooking us dinner. When he told us Mama had gone away for a while to get some rest, I figured she was tired out from running the Hideaway, which was open six nights a week till all hours. But Daddy Robert explained, "Your Mama, Miss Elaine, she been feeling down. She needed to go rest for a while."

After Mama had been away resting for about two weeks, Daddy Robert took us to visit her. He told us to put on our Sunday church clothes, even though it was a Saturday. When we got there, they told us that kids weren't allowed inside the locked part of the building where Mama was resting. So Daddy Robert went in alone to see her, while Bridgette and I stood in the hallway holding hands. The locked door had a window with wire crisscrossed inside the glass, like it was a cage. Or maybe the wires were an electric force field to protect the people inside—some sort of shield against whatever it was that broke them down in the first place.

"Where are we?" I asked Bridgette. "Is this a prison?"

"This is a hospital where they bring crazy people," she whispered, squeezing my hand tighter. I always knew Mama sometimes acted crazy. But I never thought she was actually a crazy person. I didn't let go of Bridgette's hand until Daddy Robert came out.

When Mama came home a week later, she went straight to bed and stayed there for two days. I asked her why she was so tired out after resting for three weeks. She said they gave her medicine for when she got sad, and the pills made her sleepy.

That's how it went with Mama. She was either tired or sad or angry—or else she was full of fun. Guessing her mood was harder than seeing through the backs of cards at bridge. And it was scary too, because if you guessed wrong you could walk into something nasty when you didn't have your force field up.

WHEN DADDY ROBERT was gone out to sea, Mama liked to have other men over to the house. I don't know if she met them at the Hideaway, or somewhere else. She never seemed sad or tired when these men were around. She was happy and sassy and laughed a lot. She'd put some music on and soon she'd be moving to it and smiling like she knew how fine she looked.

When a man came over who wasn't Daddy Robert, Bridgette got all scowly and would try to take me back to our bedroom, even if it was daytime. But I didn't like being cooped up back there, and I wanted to see who Mama was hanging out with. Sometimes her boyfriends would give me chewing gum, or some change from their pockets. But mostly, they were interested in Mama.

For a while Mama was dating Mr. Henry, who rode a motorcycle and always wore a leather jacket, even inside. One day I saw them on the couch kissing each other with their mouths open and their tongues touching, like I'd never seen folks do before.

A week later when Daddy Robert came back to home port, he dropped his duffel bag by the door and kissed Mama hello all sweet and gentle on the cheek.

"You don't kiss Mama like Mr. Henry do," I said.

Daddy Robert looked at Mama all fishy-faced.

"Don't pay that crazy boy no mind," Mama said, putting her arms around Daddy Robert's waist and squeezing him close. She glared over his shoulder at me with a look that made me want to run and hide. I couldn't help it. Whatever came into my head just popped outta my mouth. It's no wonder I was always getting my ass whipped.

Bridgette grabbed my hand hard and pulled me toward the bedroom. "He just got a vivid imagination, Daddy Robert," she said. "Don't pay him no mind."

I don't know if what I said was the cause of it, but pretty soon we were packing up the Maverick and driving out of Houston, leaving the Hideaway and Daddy Robert in the rearview mirror. Bridgette was super sad to leave Daddy Robert behind. She said he was like a real daddy to her. I mostly missed the stuffed peppers he made us and the way his gold tooth flashed when he smiled. What made me sad was not getting to say goodbye to Bobby and his folks.

But that's the way it went with Mama. She was a mover, and when it was time to pack up and go, we were gone.

5

SOMETIMES A GOOD THING can turn into something bad.
When I turned eight, Mama gave me what I wanted most in
the world for my birthday: a black BMX bike with high-up handlebars. It even had a license plate mounted behind the banana seat
with the number 8 on it. By then, we were living in Pomona, near a
GM parts plant south of L.A., where Mama had found a job. Our
neighborhood, known as Patty Track, was controlled by a Latino
gang called West Side Pomona. The hood next door was call Sin
Town, which was run by a Black gang called the Sin Town Crips.

Naturally, I wanted to ride my new BMX to school, and naturally,
I became an instant target. Each morning, Bridgette walked me
across Patty Track and Sin Town to school. But once inside the
school gates, I was on my own. During recess, the kids would gather
up into gangs. They didn't belong to real gangs yet, but depending
on their neighborhood and whether they were Black or Latino, the
kids would join blacktop gangs with made-up names like the OG
Crips or West Side Mafia—just so they didn't get messed with. But
not me. I was a stay-to-myself kid who never felt like joining groups.

There was a scary guy in the class ahead of me who was always
shaking down younger kids for lunches and candy. We called him
Now and Later, after the two-toned taffy candy everyone was into
that year and that he was always taking off us smaller kids. Now and
Later was a big hulking dude who wore XXXL sweatshirts to make

himself look even bigger. He must have been ten or eleven years old, even though he was still in third grade. Now and Later almost never had to resort to violence. He just intimidated the hell out of us.

Sure enough, the very first day I showed up at school with my bike, Now and Later homed in on me like a laser. After school let out, while I was practicing figure eights on the blacktop, I spotted him watching me from over by the rusted-out jungle gym. Now and Later didn't even come up to me. He just motioned me over with his right hand. I pedaled in his direction, careful not to stop as I made wide circles around him.

"Yo, punk," he called to me. "They tol' me you was talking 'bout my mama."

"No. Not me, man. I wouldn't talk about yo' mama."

"Gimme the bike," he said. He kept his body dead still like one of those king cobras that stares you half to death before he actually kills you.

I kept pedaling in wide circles. I glanced across the blacktop at the exit gate, wondering if I should make a break for it. It's not that I hadn't learned how to fight by then. Every time I showed up at a new school or on a new block, I'd get tested. Usually on the first day. It wasn't just slap-boxing or trash-talking, but real fist-fighting. I could handle myself with kids my own size. But I hadn't figured out how to work around the big guys like Now and Later.

Seeing me hesitate, Now and Later lowered his voice to where I could barely hear him. "Give it up, punk. That's my bike now."

My legs were shaking by then. I got off my bike and leaned it up against the jungle gym. He swung one leg over the seat and rode lazily out through the gate, standing on the pedals. He looked like the big bear riding a tiny bike in the circus. I tried not to cry as I watched my number-8 license plate disappear out of sight.

As scared as I had been of Now and Later, I was terrified of what Mama would do to me when she heard what happened to my new bike. Outside on the streets, the gangs were in charge. Inside our house, Mama ruled.

When I told her about Now and Later, Mama thumped her temple with the heel of her hand. "What you tellin' me? You let a boy steal your bike offa you, and you ain't even try to kick his ass? You just gave it away?"

"But Now and Later ain't just some boy, Mama. He the biggest and baddest bully at school. Don't nobody mess with him. He scary, Mama!"

"I'll show you scary!" She leaned over and grabbed my orange Hot Wheels tracks off the living-room floor. Cars went flying in every direction, and I tried to dive under the couch. I could only get halfway under it before I felt the tracks coming down hard on the back of my legs. It hurt worse than a belt. "You betta go get that bike back," she muttered, tossing aside the Hot Wheels tracks, "or I'ma really beat yo' ass."

FIRST THING I DID was scour the neighborhood. Just maybe, I prayed, Now and Later dumped it somewhere so he wouldn't have to explain to anyone back home where he got a brand-new bike. I stayed out till past dark looking behind bushes and dumpsters and abandoned buildings. No dice.

I started to panic. I couldn't go home empty-handed, and I couldn't stay out all night on the street. So I moved up the chain to my cousins, the ones who practiced their kicks and chops on me and were now wearing the blue of the Grape Street Watts Crips. I didn't really want to turn to them because, well, I didn't want nobody killed. These guys didn't just play at hard—they took care of business. I could be starting something that I didn't want them to finish. But I was up against it and scared to death of facing Mama without my bike.

My cousins got my description of the bike and of Now and Later, and put the word out on the street. An hour later, three of them came to get me. "C'mon, li'l cuz. Time to repo that bike o' yours."

Now and Later turned out to live on the worst block in Sin Town. The front door to his apartment building had no doorknob or lock, as if the police had rammed it in. The overhead light in the hallway flickered on and off and made everything look spooky. Bad smells and creepy sounds came through the doors of open apartments. I just wanted to run out of there, but my cousins rolled through the hallway like they owned the joint.

When they banged on the apartment door, it creaked open on a half-broken hinge. They walked in, three abreast. I shuffled in behind them, just hoping Now and Later wasn't there.

I saw him slumped on the sofa next to some grown-ups watching a basketball game on a big console TV that had a jagged zigzag pattern running across the screen. Piles of clothes and other junk were heaped in the corner and spread across the floor. Every horizontal surface had an overflowing ashtray of cigarette butts. No one looked at us or got up to say hey.

Mama used to tell me and Bridgette, "No matter how bad it gets for us, we ain't nearly as bad off as some o' those other folks." Standing there in Now and Later's living room, I thought to myself, These are those other folks. Everything about that room and the people in it just cried out "bad" and "sad."

My oldest cousin shouted over the basketball game. "We came for my li'l homey's bike." After a few seconds, one of the grown-ups on the couch grunted and jerked his head toward the far corner of the room. There was my bike, lying in a heap. It was beat down like some broken bird, the handlebars and seat twisted sideways as if someone had jumped up and down on it till it played dead.

One of my cousins hoisted the bike on his shoulder and we headed for the door. As we were leaving, I sneaked a peek over at Now and Later. He had shifted down so low on the couch I could barely see him. Just then I felt sorry for him and for the sad, broken-down home he had to come back to every night.

My bike cleaned up okay. The license plate was gone—but I wouldn't be eight forever. My BMX looked sort of lived-in now, which was just as well. A shiny new bike brings out the mean in folks. It's like what they say back in New Orleans about crabs in a bucket: as soon as a crab tries to crawl out of the bucket, some other crab will sho'nuff grab ahold of him and pull him back down. Humans have the same mentality, as I was learning. Anytime you try to raise yourself up, someone with less is gonna try to grab on and pull you down.

6

THE FIRST TIME I remember feeling poor was in third grade. We were living with the Brown family on Curran Street, back in New Orleans East—which is where we moved to after Pomona. Mama's old friend Miss Jeannie worked steady at the post office. But raising up her seven kids on government wages meant there was never quite enough to go around. Meanwhile, Mama was having trouble finding work in New Orleans, and she was way too proud to go on food stamps or welfare.

I figured out we were poor because I was hungry a lot of the time. We didn't starve in those days so much as eat simple. Fast food was a luxury we couldn't afford. Mickey D's was strictly for birthdays. To get by day to day, we'd buy food that could stretch. At the Browns' it was grits and margarine for breakfast every morning, and for dinner we had red beans and rice several nights a week.

If you could find an egg—and sometimes you could if you knew where one was hidden in the house—you could make a toad in a hole by cutting out a circle in the bread with the top of a juice glass and frying the egg inside. If we were craving something sweet, we'd make huck-a-bucks by pouring Kool-Aid into paper cups and freezing them. Then we'd turn them upside down in the cup and suck the sweet juice outta the ice.

When the house was empty of food, we'd put mayo or ketchup on bread and call it a sandwich. Syrup sandwiches were my favorite.

When there was nothing else left to eat, we'd eat balled-up bread, which tasted like more than just bread. But there was no getting around the math of it. In a house with ten mouths, being the youngest meant a hunger hole in my belly most of the time. The free lunch at school was the only meal I could count on.

One day when we were out of food and it was still a week away from payday at the post office, Miss Jeannie turned to Mama and said, "We gotta break the bank."

"Yeah," Mama sighed. "I guess we really gotta."

Mama loved that bank. It sat next to the TV console, underneath a picture of a white Jesus. It was a large clay coin bank, but instead of looking like a pink piggy it was a black clay sculpture of a naked couple kissing and hugging each other. Whenever Mama and Miss Jeannie came into the house, they deposited their loose change through a slot in the back of the woman's neck. There wasn't any other way to get inside that bank. I knew because I spent a lot of time looking for a secret hatch on the side or the bottom.

Everyone gathered around while Miss Jeannie spread a sheet on the floor of the living room. Mama carried the bank over with both hands and set it down in the middle of the sheet. I couldn't wait to find out where the secret door into the bank was hidden. When Miss Jeannie came out of the kitchen holding a hammer, I was horrified. It never occurred to me that "break the bank" meant breaking the bank for real. That's when I knew we must really be broke.

Miss Jeannie raised the hammer overhead while Mama turned away like a pet animal was about to get shot or something. The hammer came down hard on the kissing couple, the bank cracked wide open, and coins and pieces of clay splattered across the floor.

"Hoo-eee!" somebody shouted. "Jackpot!"

We all scurried around the living room collecting coins and piling them in the middle of the sheet. It made a serious pile. I had never seen so much money in my life. I was already counting the coins with my eyes, when Mama said, "Hey, Bionic Brain, set you' ass down here and do your thing!"

I was surprised, because Mama said it sort of proud, like she was bragging on me and my special counting power. Praise and approval were as scarce as money in our household. I was used to Mama yelling at me to stop walking with my head down and counting my steps all the time. She was happy to finally put my counting powers to good use, and so was I.

Everybody stood around in a circle while I sorted the coins into stacks and counted them up in my head. In less than three minutes I announced the total: "One hundred and sixty-four dollars and seventy-four cents!" I looked over at Bridgette, who was smiling like she was proud of me too.

THE OTHER TIMES I got proud looks outta Mama were when we did puzzles together. Going all the way back to Houston days, Mama liked to relax by lighting a joint and curling up on the couch with her Dell puzzle book. She was always happy to have me join her on the couch and help her out.

Sometimes we'd work side by side, with Mama doing a crossword while I did a Kriss Kross. My specialty talent, though, was cracking the logic puzzles. Here's the kind of story problem you'd have to solve in a logic puzzle:

A group of students decided to form a drum corps. One is the drum majorette, one is the bass drummer, one is the side drummer, and the remaining two are the cymbalist and trumpeter. No two drum-corps members are the same age. From the information given, determine each corps member's name, age (14, 15, 16, 17, or 18), and position. (Note: Young women are Amanda and Esther, and young men are Leonard, Mark, and Owen.)

Mama couldn't solve logic puzzles like that one. But I could. And fast! I'd stare at the page without blinking for about a minute, and then I'd shout out the answer all in a rush: "Amanda, eighteen, drum majorette, Esther, seventeen, side drummer, Leonard, fifteen, bass drummer, Mark, fourteen, trumpeter, Owen, sixteen, cymbalist!" Then Mama would turn the page and see that I'd gotten it right. She'd shake her head in amazement while I did a little happy dance in front of the couch, busting my best moves and making Mama laugh.

7

Becoming best friends with Darin Brown, the youngest of the seven Brown kids, was my first clue that life could be more than a pile of crabs trying to pull you back into the bucket. Darin was the guy who reached out from the crowd of bullies and doubters and lifted me up.

Darin had a square head with a chipped front tooth and he often made a squinty face, like when someone is looking into the sun. He sure didn't look like the angels in the big leather-bound King James Bible that sat on our coffee table—the glowing white ones with long blond hair who swooped down from heaven to save Isaac from his daddy's knife or Daniel from the lions. But I definitely felt like the miserable Israelite stuck in a pit of danger, and Darin was my guardian angel.

Now that Bridgette was a teenager and was sleeping with the Brown girls in the room across the hall, Darin was the one who watched over me. He was just two years older and three inches taller than me, but Darin became my shield. I was still the youngest and smallest in our crew, and a magnet for bullies. Anytime something started up, in the house or on the street, Darin would step in front of me. Once, he even caught a rock in the face for me.

Darin stood out from the rest of the Brown family crew like a shiny penny. His six brothers and sisters had three different daddies. One of the daddies was doing hard time in Angola Prison, north of

Baton Rouge, and Darin's two teenage brothers were hard dudes who seemed to be on a fast track to Angola themselves. His older sister could be just as nasty as they were. We called her the Hammer because Miss Jeannie and Mama put her in charge of keeping the rest of us in line. She whipped us more often than our mamas did.

Darin was different. To me, he embodied the virtues of our favorite cartoon heroes: the honor and integrity of Aquaman and the intelligence of Peabody in *Rocky and Bullwinkle*. He was a natural leader and a great athlete. He was my intellectual companion and inspiration, the person who made me feel good about being smart, instead of just weird and different. Most important, he was kind. Darin was the first guy who acted like I was someone worth protecting.

Darin and I did all the best-friend things together. We played touch football in the street every day. We explored the canals and woods on the edges of New Orleans East. We made some mischief together, like sneaking into swimming pools at housing complexes, until we got caught and kicked out.

Inside the house we played board games, cards, and make-believe. We were both interested in animals. Our favorite TV shows, which we watched together every week, were *The Undersea World of Jacques Cousteau* and *Wild America*. Darin gave me someone to compete with. We pushed each other to be faster, smarter, funnier. And best of all, Darin was the person who taught me how to get books.

EVER SINCE I TURNED SEVEN, I'd been reading everything I could get my hands on. Which wasn't much. The only reading material in the house, besides the Bible, was Mama's *Reader's Digest*. I was hungry for more. And the only public library in our neighborhood had shut down and moved across the canal.

Darin came to my rescue. He figured out that you could join the

Book of the Month Club just by cutting out the ad in *Reader's Digest,* checking the "Bill me later" box, and mailing in your membership. Darin read me the tiny print at the bottom of the ad that said "You must be 18 years old to enter into this contract." Since I was just a kid, it wouldn't be a real contract, he explained, so they couldn't send me to jail if I didn't pay.

So I joined the club and started receiving books in the mail every month. While everyone else was watching TV or playing bid whist, I was hiding out in the bedroom with my head inside of Nancy Drew novels and scary-ass books of ghost stories like *The Goblins Will Get Ya!* Then I moved up to Edgar Allan Poe short stories. After a while, the Book of the Month Club got tired of not getting paid, and stopped sending me books. So I joined the Time-Life Book Club and ordered books about nature, animals, and the ancient world.

The book that really uncorked my mind was Edith Hamilton's *Mythology.* I picked it as my April selection because the cover illustration looked like a superhero comic book—a guy in a toga riding a winged horse across the sky while shooting golden arrows from his bow.

The Greek heroes were actually a lot like comic-book superheroes with superpowers. But they had to have a god on their side or they were screwed. In Greek mythology, getting in good with the gods was even more important than having a powerful mama or daddy. The gods could protect your ass if you were their favorite, or they could make you miserable if you crossed them. They reminded me of the all-powerful grown-ups in my life.

If a mortal broke their rules or stole from them, the Greek gods acted like gangstas who wanted to dream up a punishment so evil that nobody would even think about crossing them again. If a guy stole fire from them, they would tie him to a rock and have a giant eagle eat out his liver every day. They made another guy push a heavy rock up a hill and watch it roll back down, day after day, forever!

There was no escaping the wrath of the gods—unless you were one of the really clever ones. The Greek heroes I admired the most weren't the musclemen like Hercules, or even the bravest warriors like Ajax and Achilles. I wanted to be like Ulysses, who used his brains to outsmart the gods. It took Ulysses ten years to make his way home from the Trojan War, and he lost his ship and all his men along the way. But he survived because he was smarter than everyone else.

But here was the really spooky thing about Greek mythology: You couldn't always tell the gods from the mortals, because the gods sometimes wanted to walk the Earth disguised as human beings, or even as animals. And even if a god liked you and gave you a superpower—like being able to see into the future, or fly like a bird—half the time it turned into a too-much-of-a-good-thing problem, like riding a shiny new bike in the ghetto.

You just never knew if the creature swooping down out of the sky was a winged horse who would carry you up to Olympus, or an eagle come to pick at your liver.

8

My winged horse arrived on a Saturday afternoon in the middle of a chess game.

Darin had taught me how to play chess right after I turned nine. Every Saturday we'd play a famous chess match the *Times-Picayune* published in the back of the Sports section, next to the bridge games. That Saturday morning the *Picayune* printed Game 6 of the 1972 Fischer-Spassky world championship, which the caption explained Fischer had won with a combination of the Queen's Gambit Declined and the Tartakower Defense. Everything in chess sounded like gang warfare, with military strategies like the Sicilian Defense, the Harrwitz Attack, and the Poisoned Pawn Variation. I liked chess because you could try to dominate and defeat your opponent, but nobody got bloodied for real.

After playing out the match move by move, Darin and I would try to replay the game without looking at the newspaper. We'd get as far as we could by memory and then we'd play out the game to the end ourselves. Darin had been playing chess longer than me. But I was better at memorizing the moves, and I could see the board better than most people. A chess board has sixty-four square spaces, but if you looked at the board another way, you could see larger squares made up of groups of smaller squares. When you added up all the different-sized squares on the board—64 + 49 + 36 + 25 + 16 + 9 + 4 + 1—there were actually 204 different squares. If you could play on all 204 squares, like I was learning to do, you could also see more moves into the future. While Darin was playing two- or three-move sequences, I was planning five or six moves ahead.

I was playing White that morning, which meant I got to spend an hour inside Bobby Fischer's game and feel the superhuman level his mind worked on. He was like a Greek hero with a fatal flaw. He was super intelligent, but being so smart drove Bobby Fischer crazy to the point where he dropped out of chess three years after he became world champion in 1972. That was just the kind of scary shit the gods liked to pull on mortals.

We were on move twenty-four when Miss Jeannie called to me from the living room. "Li'l Jame, c'mon in here." I ignored her, because I was about to pin Darin's rook, and I wanted to make sure I wasn't stepping into some trap he'd laid for me.

"LI'L JAME!" she hollered. "Don't make me come up and drag yo' ass outta there."

We got up from our game and wandered out to where a group of grown-ups was sitting around the table drinking iced tea and smoking cigarettes. There was Mama and Miss Jeannie and some guys and a woman I didn't know. Darin and I sat down on the edge of the couch across from them, but we didn't settle in because we wanted to get back to our game.

"You know who that is?" Miss Jeannie asked, pointing at a man with a beard and a wide-brimmed hat sitting at the table. "Go sit on his lap."

"C'mere, boy," the man said, smiling. "I ain't gon' bite ya." I remembered that voice. I went over and sat up on his knee like I was told. When I looked out at the grown-ups, they were all grinning back at me.

"That's yo' daddy!" said Miss Jeannie, giving out a hoot in that high-pitched laugh she had.

I looked over at Mama, who didn't say anything. She just blew some lazy round smoke rings and looked at us with a mysterious smile on her face. She reminded me of the spooky hookah-smoking caterpillar in *Alice in Wonderland* who asked Alice, "Who are you?" I couldn't read Mama's mood, which was nothing new. But her weird smile put me on notice that something wasn't normal—that things were about to change.

To be honest, I hadn't thought much about my daddy during the years we'd been out west. Mama rarely talked about him. He never called or showed up on special occasions. No one I knew had a daddy around, so it seemed normal that mine was a ghost who I barely remembered except for his voice, and that night of the burning bed.

But now he was right here, close enough that I could smell the sweat and weed and men's cologne on him. I twisted my head around to look at him, to see if I could find my face in his. He had light skin and gray eyes. My eyes and skin were both darker. He was a good-looking, fit man with a fine Afro—nothing like my nappy head of hair—and he had an easy, confident manner. I was this dark, crooked-toothed boy who couldn't ever sit still. Maybe he wasn't my daddy after all.

Soon enough, the grown-ups went back to grown-up talk and ignored me and Darin. We sat on the couch and watched the grown-ups. I couldn't take my eyes off of Daddy. I just wanted to study him. I noticed how when he smiled or laughed, everyone else smiled and laughed too. Even Mama, who barely ever smiled and rarely laughed out loud. He had a singsongy voice that was different from the way most people talked in New Orleans. He was one cool dude, and I could tell that people liked him.

When Daddy got up to leave, he turned and pointed to me and said, "I'm gonna take you home to Piney Woods. Real soon."

When I asked Mama about Piney Woods, she just shook her head. "That's the country." I could tell by the way she said it that she didn't have any use for the country. She was New Orleans East, and proud of it.

9

THE NEXT SATURDAY, just like he said he would, my daddy picked me up to take me to visit Piney Woods. I got to ride up front alongside him in the pickup truck. As we were heading out of town, Daddy stopped at a delicatessen at the corner of Dale Street and Chef Menteur Highway. He got out of the truck and motioned me to come with him into the store. He walked right up to the deli counter and started ordering up food. Real food. Didn't even ask what things cost. He bought us big hunks of bologna and salami and hogshead cheese, plus a box of saltines.

I could tell Daddy knew the guy behind the counter because of how they were talking and joking. The counterman asked him who I was.

"Tell 'em your name, boy," Daddy said to me.

"I'm James Plummer Jr.," I said. That cracked Daddy up, just hearing me say my name. Mama, Bridgette, and Miss Jeannie called me Li'l Jame, but everyone else called me Plummer or James Plummer Jr. It had somehow never occurred to me that my daddy was the *original* James Plummer.

We got back in the truck and started eating right there in the front seat, using a Buck knife Daddy kept in a sheath on his belt to cut up the food. It was a banquet for a boy used to going hungry or eating grits and margarine twice a day. "Slow down, James Plummer Jr.," Daddy said. "You gonna bust a hole in your belly if you don't."

While we got onto Route 59 North, I found a map in the glove compartment and figured out that it was 161 miles up the highway to Kelly Hill, which was in central Mississippi. I calculated it would be about a three-hour drive. I wanted to get on Daddy's good side so he'd want to tell me stuff about himself. I started by telling him jokes. Silly stuff I'd heard and memorized from TV shows like *Welcome Back, Kotter*. When he laughed at a joke I'd tell him more of the same kind. Then I'd sneak in a question about him and his family. I was counting the mile markers along the highway to make sure I got in as many questions as I could in between the jokes. By mile 150, I'd filled in a lot of blanks.

Daddy told me he grew up poor and country in backwoods Mississippi. He was born in Pachuta, which wasn't actually a town, just the closest post office. The official name of the area was East Barnett, but folks who lived there called it Piney Woods, since there were mostly pine trees growing in those parts. When Daddy said he was born in 1933, I pitched it back to him real quick that he must be forty-three. "Forty-three's a prime number," I added.

"Elaine told me you has a brain on you. She right about dat."

He told me there had always been three families in Piney Woods: the Plummers, the Stricklands, and the Kellys. They all prided themselves on having light skin, light eyes, and "good hair." The reason they were light is there was lots of mixing back in the old times. My daddy had two white Irish grandfathers, one on either side. His grandfather on his daddy's side had two families, a white family and a Black family. When he died, he shocked everyone by leaving a large piece of land in Clarke County to his Black family, even though they didn't have his name.

The Plummers, Stricklands, and Kellys all shared Piney Woods, and mostly they got along. But they could be fierce and competitive too, and fights would break out, some of them notable. Daddy told me they even had a fight once about which family was prettier that ended up with some serious blood being spilled on all sides.

Daddy was the second youngest of eleven kids, and he dropped out of school at nine to work the farm. He left home at thirteen and went to New Orleans, where he got a job parking cars for twenty-five cents a day. When he turned eighteen, he joined the army and went to fight in the Korean War. After he was discharged, he stopped over for a while in Alaska, where he boxed as a middleweight. Then he moved back to New Orleans and landed a job at the Kaiser Aluminum plant that had just opened in St. Bernard Parish. It was hot, nasty work in the furnace room, and he had to eat salt tablets all day just to keep from passing out from dehydration. But Kaiser paid better than most other blue-collar jobs in the area.

Daddy was still working there now, twenty years later, but he always had some side hustles going, including on Kelly Hill. The Plummers were the most entrepreneurial family in Piney Woods. They became bootleggers back in the '20s, and when Clarke County stayed dry after Prohibition ended, they kept that business going.

JUST AS WE TURNED OFF Route 59 toward Kelly Hill, Daddy shouted, "Looky there!" and pointed at an animal that was rooting in the dirt by the side of the road. I'd seen pictures of armadillos in books and on *Wild Kingdom,* but I'd never actually seen one up close. Daddy swung the pickup onto the shoulder of the road and grabbed the .22 rifle from the rack behind our heads. He stood by the driver's door, aimed at the armadillo, and fired. The armadillo jumped about four feet straight up, somersaulting like an Olympic diver, then landed and scurried into the woods.

"C'mon, boy," Daddy shouted, "let's go!" Daddy ran into the woods with his rifle held high in one hand. I scrambled out of the pickup and followed behind him.

Other than a few parks and patches of trees outside New Orleans, I'd never been in the woods before. I kept stumbling on roots and vines and bruising my shins on downed trees. But I bounced up and kept on running so I wouldn't miss out on the hunt. Daddy was sprinting, almost dancing, through the woods, like

O. J. Simpson running through the secondary. Then suddenly he stopped, leveled his gun, and fired. By the time I caught up to him, he was crouched over the dead armadillo and had pulled his big Buck knife out of the leather sheath on his belt. He flipped the dead animal onto its back so its belly was exposed, then he plunged the blade in high up and split that animal from head to tail. It turned out to be a girl armadillo, because when he dumped its insides onto the ground, there were four tiny little armadillos, with no armor whatsoever, wriggling there among the guts.

"Oh, hell," said Daddy shaking his head. "Don't you be lookin' at that."

He dug a little hole with his knife in the ground, pushed the tiny critters into the hole with his boot, and covered them over with dirt. Then he hoisted the gutted armadillo by the tail in one hand and his rifle in the other and led the way back to the pickup.

10

"WE BROUGHT DINNER!" Daddy shouted out to Aunt Middy, who was sitting on her front porch on top of an overturned tin tub, shelling peas.

Aunt Middy was pretty much the fattest woman I'd ever seen. "Large and in charge," is how Daddy described her on the drive up to her house on Kelly Hill. She ran the family farm, and the family enterprises in Piney Woods. Aunt Middy was Daddy's sister-in-law, and his business partner.

While she looked me over, I watched in amazement as she dug her finger into her lower lip and pulled out a moist packed wad of a brown powder. She immediately replaced it with a large pinch of dry brown powder from a corked glass jar. Then, she pressed her index and middle finger up to her lips and spit a stream of brown juice into a narrow-necked jug beside her. All the while she kept her eyes fixed on mine.

Aunt Middy lived in the bigger of her two houses on Kelly Hill, along with her two brothers, Uncle Rosie and Uncle Henry; her hired hand, Mr. Will; and Andrea, who turned out to be my sixteen-year-old half-sister. My daddy was her daddy too. Andrea's mama—who grew up with my daddy on Kelly Hill—had moved to Detroit when Andrea was still a baby. Aunt Middy, who didn't have any children of her own, had raised her up.

Andrea took the armadillo from Daddy and I followed her inside, where she threw it onto the red-hot coals in the fireplace. When it had charred up good, she pulled it out of the fire by its tail and scaled it with a knife, like you would a fish. Then she took a sharp machete and chopped it into small pieces, armor and all, and threw them into a big steel pot. She added some chopped onions, green beans, and potatoes and lifted the pot onto the coals.

"You met your brothers yet?" Andrea asked me. That's how I learned about Byron and Fionne, Daddy's two sons who lived in New Orleans with their mother, Miss Carrie. It turned out that Daddy had Byron with Miss Carrie a few years *before* he met Mama. After Mama set their bed on fire, Daddy married Miss Carrie and had Fionne, who was four years younger than me. Andrea told me I had another sister in New Orleans named Yolanda, but I couldn't figure out which branch of the family she was connected to or who her mama was.

I asked Andrea if she had a pencil and paper I could borrow, so I could draw myself a picture of my fast-growing family tree—which was suddenly feeling more like a blackberry bramble, all twisty and thorny. At the center of this briar patch sat my daddy, who was starting to remind me of Zeus with all these kids and mamas scattered across three states. And those were only the brothers and sisters I knew about.

AUNT MIDDY'S HOUSE had electricity, but no plumbing. You had to go out back to draw water from a pump and to use the outhouse. The house wasn't fancy, but the Kelly Hill farm could feed the whole extended family. During our armadillo dinner—which Daddy declared tasted like fine barbecued pork—I learned that Aunt Middy had geese, ducks, guinea hens, setting chickens, cows, pigs, and hogs, plus a three-acre vegetable farm on the side of the hill and a mule to plow it with.

Before bed, I needed to pee, so I went out back. It was a moonless night, and scary dark like I'd never seen in the city. So dark that

I stumbled over some gardening tools and fell to the ground. Even the dirt smelled different—like it was alive. Everything about Kelly Hill was strange and spooky. The woods. The food. The animal sounds and smells. Even the way people talked sounded distorted, like a record playing backward.

Very slowly, I started to feel my way through the dark toward the outhouse, counting my steps so I could retrace them back to the house. I heard a voice inside my head—was it Daddy's?—say, "Look up, boy!" So I did.

The night sky looked like someone had punched a hundred thousand holes in it, then shined a bright light from behind. It confused me the way the stars looked so far away but at the same time bright and close enough to reach up and grab with your hands. They made me feel tiny and huge all at once—the same way I felt watching Daddy jump out of the car and shoot that armadillo and then gut it, one-two-three.

I just stood there and stared up at that sky, full to bursting with stars. I couldn't imagine how to count them all. But I knew I wanted to.

11

THE FIRST LESSON I learned about country life was that everyone gets up early. Andrea shook me awake at dawn the next morning to help her empty the nighttime piss pots, slop the hogs, and feed the chickens.

As soon as we'd finished our flapjack and 'cane molasses breakfast, Daddy and I walked up the hill to see how the summer crops were coming in. It looked just like a farm in a kid's picture book, right down to the crucified scarecrows. Daddy left his shirt in the truck while we walked the fields. His torso was strong and muscular—I guessed from growing up working hard on the farm, and then spending twenty years hoisting vats of aluminum in the furnace room at Kaiser Aluminum. Watching him walk the rows of tomatoes and cucumbers and chicory, I could imagine him as a young boxer, fighting in the smoky back rooms of Seward and Fairbanks. He looked like a Greek demigod—like Perseus or Hercules. And he was my daddy, which made me a quarter-god!

When Daddy disappeared between two tall rows of corn, I followed him into the field. Between each row grew a line of bushy, weedy plants. It wasn't until I watched Daddy inspecting the plants closely that I realized that the weeds were . . . weed! I'd seen plenty of folks smoking weed at card parties and on street corners, and

Mama liked to unwind with a joint before bedtime. But I'd never seen it growing out of the ground before.

Andrea was the one who clued me in to the family business. Aunt Middy sold nickel and dime bags of weed out of her house, while Daddy took big sacks of it back to New Orleans to sell. And since Clarke County was dry, Aunt Middy trucked in booze from Meridian to sell out of the house. During weekend softball games she and Uncle Henry sold beer out of the back of her pickup.

THAT AFTERNOON, Daddy took me for a drive around Piney Woods and then headed down Highway 11 toward Laurel, which was the closest thing to a city in our part of Mississippi. As soon as we crossed the line into Jasper County, he pulled his pickup into a parking lot in front of a small unmarked building. It wasn't much bigger than a shack and it didn't have a sign or anything, though Daddy referred to it as the B&M Club. Inside was a pool table, a bar, and a small dance floor that opened up to a patio out back. There weren't any customers inside on a sunny Sunday afternoon, but you could see from the empty beer bottles and overflowing ashtrays that there must have been a full house the night before.

When we walked inside the club, Bobby Kelly came out from behind the bar to give Daddy a bear hug. Bobby popped the top off a beer and set it down on the bar in front of Daddy.

"Tell Cousin Bobby your name, boy," Daddy said to me, already laughing at what he knew I would say. Bobby opened a root beer and handed it to me while I did my "I'm James Plummer Jr." bit. I did love to hear my daddy laugh.

As he steered his pickup back toward Piney Woods, Daddy explained that he and Aunt Middy were buying the B&M Club off of Cousin Bobby. Mama would run it. Bridgette and Andrea could work there too. Mama and Bridgette and I would move into Aunt

Middy's house—at least until we could find or build a house of our own.

"You'll go to Quitman Elementary," Daddy said, clapping one of his strong hands on my shoulder and squeezing it hard. "Same as where your daddy went."

So just like that, I went country.

12

You WOULDN'T THINK that backwoods Mississippi would turn out to be a training ground for a life as a research scientist. But that's where I learned my work-hard habits that would power my journey to a galaxy far, far away from Aunt Middy's farm.

Bridgette and Mama and I moved into Aunt Middy's house in July. But summer wasn't a lay-around, play-around season on Kelly Hill. Every day on the farm felt like a competition to see who could work the hardest. Everyone was part of the extended family, and everyone worked to support one of the family enterprises—either the farm, the weed-and-booze business, the B&M Club, or some other side hustle to bring in extra cash, like hauling pulpwood or roofing or bricklaying.

We were all expected to get up early every morning and get at it. Rising before dawn to work was the country way to beat the heat. Even my new roommate, Uncle Henry, a cranky old wino who woke up with both hands shaking, put in a full morning's work hoeing, plowing, chopping wood, and castrating pigs before knocking off in the afternoon to begin drinking.

"You best feed dem chickens and hogs if you 'spect ta get fed in dis house," Aunt Middy told me the first night we moved in. So after a flapjack breakfast, I headed out to the chicken yard behind the house. As soon as I scattered the seed the yard was teeming with chickens. A flock of a dozen black guineas with white faces swooped in from around the barn. Then a rooster chased down a hen and jumped on her back, grabbing the feathers on her neck with its beak. The rooster's wings folded down around them while he did his business. A few seconds later, he jumped off and walked away. The hen climbed back to her feet, unruffled her feathers, and went on eating seed like nothing had happened.

I wasn't sure what I'd just witnessed, but I wondered if it had something to do with eggs. I found a shady spot alongside the smokehouse where I could sort that out. Suddenly, I felt something burning and stinging my feet. When I saw that I was standing in a swarm of red ants, I jumped up and ran back to the house as fast as I could fly.

"Don't you go bringing none of them creeping-crawling critters into this house!" Mama shouted, shooing me onto the porch to nurse my bites with an ice chip that melted in about thirty seconds in the morning heat.

Lesson number three of country living: Everything is alive and reaching out to grab you, sting you, bite you, or poison you. I'd seen plenty of mosquitos and roaches back in New Orleans. But those city critters wouldn't have survived five minutes on Kelly Hill.

As I got to know the farm and the surrounding woods, I was introduced to a host of attacking plants and pests. Red ants lurked underfoot with pincers on their heads and a stinger on their rear ends. Horseflies terrorized me by day and invaded the house at night. Someone would yell, "Horsefly in the house!" and everyone would be after it. No one could go to sleep until the horsefly was hunted down and killed. Otherwise it would feed on us while we slept, leaving incredibly painful welts.

Cocklebur bushes were covered in little spikes that could hurt something fierce if you accidentally stepped on one. And if you stumbled into a bramble patch in the forest, it could take you half an hour to fight your way out. Worst of all was the poison ivy and poison oak. One touch and a nasty rash would break out and spread across your body.

It seemed like every day I was running to Mama or to Aunt Middy with a sting or a bite or a rash that set me howling. Aunt Middy instructed Mama to chew up a wad of tobacco and drip the juice out of her mouth onto whatever welt was rising off of my body. Mama thought chewing tobacco was low-down country behavior, but tobacco juice was Aunt Middy's all-purpose medicine, and no one went against Aunt Middy in her house.

Between the biting bugs and poisonous plants my skin got to looking something awful. I was always scratching at some bite or welt, which drove the other folks in the house into a rage. "Good Lord, Li'l Jame, you always scratchin'!" cried Mama or Bridgette or Andrea or Aunt Middy. "Can't you stop scratchin' for one God-almighty minute?"

But God had forsaken me, like Job. So I'd search out a private spot in the house where I could scratch away at myself and think of how much I missed Darin. I wondered if he'd found a new friend to play chess and touch football and watch *Wild Kingdom* with. He'd never believe the wild kingdom I had to battle here on Kelly Hill.

Country life was damned uncomfortable, and no less of a fight for survival than living in the city.

13

COULDN'T HANDLE working in the fields on account of my supersensitive skin. So Aunt Middy gave me home chores to do: empty the piss pots in the morning, get water from the pump, then bring in wood to make the fire to boil the water. And that was all before breakfast! Then I had to feed the chickens and slop the hogs.

The one indoor chore I looked forward to was cleaning and bagging weed. After the weed was harvested and hung up to dry in the smokehouse, we'd dump it into big sacks, then bring it inside and pour it into large pink plastic washtubs. The weed was full of stalks and stems and seeds, so we'd all sit around in the living room after dinner, listening to music while cleaning and measuring the weed into nickel and dime bags.

I'd seen Mama doing it all my life, but she was just cleaning enough for a joint or two. Andrea showed me how to clean it a fistful at a time, and I took to it naturally. I'd sit with the latest Ohio Players double album open on my lap and propped up at just the right angle to let the weed stay in place and the seeds and stems slide down the cardboard. I felt as cool as Sugarfoot, the Ohio Players' lead singer, with his big Afro playing his double-neck guitar he nicknamed "Messiah." Every Ohio Players album had a different sexy woman on the front. The hottest one was *Honey,* which had just come out. It featured a buck-naked, caramel-skinned woman hold-ing a jar of honey in one hand while the other hand held a spoonful of honey dripping into her upturned mouth. On the inside of the double album, the same naked woman leaned backward with her whole body smeared in honey.

Using a playing card, I'd push the weed up the slope of her re-clined body, and the seeds would roll down her chest and belly and legs. Once I had a good-sized pile of cleaned weed, I'd measure it into baggies using a Diamond Red Top matchbox. One matchbox was a nickel bag, two was a dime bag. Daddy would make ounce bags that only he sold, each one filled to three of his fingers high when it was laid flat. While we cleaned and bagged the weed, the Ohio Players would sing along on the stereo:

Honey child, you got a sting like a bee girl
Oh honey, honey, honey sweet loving thing

When we'd finished cleaning and measuring a whole box full of bags, I'd separate the seeds from the stems and put them in a jar for planting. The windowsills of the kitchen were lined with seedlings in little pots, their slender stalks bending toward the sun. When the seedlings got to a certain height, Andrea would take me out to the cornfield to show me how to plant them.

THE OTHER AISLE of the family store was booze we brought in from Meridian, which was almost an hour up Highway 11, just across the county line. Aunt Middy and Andrea and I would drive there once a week in their white pickup and load up on supplies for the home store, and for the B&M Club. We sold three kinds of booze out of the house: half-pints of Canadian Mist whiskey and Seagram's Gin—called "Bumpy Head" because of its bumpy glass bottle—plus T.J. Swann, a sweet and cheap wine that came in twist-cap quart bottles in four different flavors: Easy Nights, Mellow Days, Stepping Out, and Magic Moments.

When folks came to cop some weed or booze, they'd usually find Aunt Middy sitting out front on her upturned tin tub with her legs splayed out wide, her spit bottle set next to her. Sometimes folks would come up to the house dragging fresh-killed animals, since Aunt Middy would take squirrels, coon, rabbit, or possum in trade for weed or booze. She wouldn't take fish though. Three squirrels were good for a nickel bag, two rabbits for a dime. I remember someone once showed up with a humongous beaver they'd killed. Aunt Middy roasted it on a spit and I played with its paddle tail for a few days until it started to stink.

If Aunt Middy had to be somewhere else on the hill, she'd give me an empty box of King Edward Invincible cigars to keep cash in, and I'd mind the store. When folks came knocking, I'd talk to them through the latched screen door.

"Hey na," I'd say. They'd look at me a little sideways, this nine-year-old kid with devil eyes who didn't look or sound country. If Bridgette had plaited my hair, they'd take me for a girl.

"Where Miss Middy, li'l girl?"

"She ain't here," I'd say in my deepest country voice, trying not to be wiseass about it. "I'm James Plummer Jr. What you want?"

"Damn, boy! You shoulda said that first!" Just mentioning my daddy's name got me instant respect.

At the end of the day Aunt Middy would empty the cash out of the cigar box into a stocking that she'd knot and stuff into the top of her dress. Aunt Middy had the biggest titties of any woman I'd ever seen. Even if her stocking was bulging with cash, it barely made a bump in her chest.

Aunt Middy kept ledgers of her businesses with Daddy. After dinner, she'd sit at the kitchen table writing in her ledger. She'd squint at the pages and say to me, "C'mere with dem young eyes of youz, Li'l Jame, and help yo' ol' Aunt Middy do her figures."

I'd add up the numbers and check her addition and subtraction, and she'd give me lemonade and a homemade cookie. Sometimes, when she was pleased with me because I made the numbers tie out, she'd rub my head like I was her favorite hunting dog—which pleased me too.

14

EVEN THOUGH I KNEW how to clean and sell weed, I was still a city kid wearing a Spider-Man T-shirt who didn't know jack shit about living on the land. My two country cousins, Bobby Strickland and Anthony Kelly, were only a couple of years older than me, but they were miles ahead when it came to country skills. They knew their way around the woods, running along the almost-invisible game paths that wound through the forest. They could wring chickens by the neck, two at a time. And they knew how to hunt rabbits and possum with a .22 or a shotgun. When it was time for Uncle Henry to cut the piglets' balls off, Bobby and Anthony would charge into the sty and drag the little piggies out by their two back feet. They could even drive! Most kids in Piney Woods were driving pickups around the back roads, and sometimes even on the paved roads, by the time they were ten or eleven. All I knew how to do was stumble into bramble patches and fire ant hills.

Luckily, my daddy headed back up to Kelly Hill every Friday after his shift ended at the Kaiser Aluminum plant in New Orleans. Once in a while he brought my half-brothers, Byron and Fionne, with him. But I was the only son who had all three names the same as Daddy's. Even if I was a scrawny city boy who could barely split ten logs before he had to quit with bleeding blisters, I was James Edward Plummer Jr. And Daddy smiled wide as the sky when I said my name out loud.

Daddy was the manliest man in Clarke County. The women were all sweet on him, so it was no surprise he had a reputation as a ladies' man. But most of all, he was respected for knowing his way around the tasks that made a farm run. He could fix a broken pump, start a dead truck, hitch a mule to a plow, or work all day shirtless in the field—even in the winter. When it was time to slaughter a hog, they waited till Daddy was back on the farm, because he knew how to butcher it right. He knew how to hunt and fish and where to look in the forest for the best blackberries in early spring.

He taught me all of that, plus how to aim and fire a rifle, and how to sharpen a knife blade to a fine enough edge to skin a rabbit, which was different from how you skin a squirrel. I never did shoot any animals—it just wasn't something I was interested in doing—but Daddy taught me how to shoot a can off a fence post and how to skin anything he hunted. I always liked opening things up and exploring them from the inside, whether it was a radio or a wasp's nest or a shot-in-the-head possum.

Daddy was totally country. There was nothing he couldn't do. One day in late August, he told my half-brothers, Byron and Fionne, and me that he was gonna make us a swimming hole directly downhill from Aunt Middy's house, in Cane Creek. He made us stand way back off the creek while he laid four sticks of dynamite into its banks. Then he lit the fuses and ran up the bank to join us—just as the dynamite sticks blasted in quick succession: Boom! Boom! Boom! Boom! When the dirt clods had all fallen to the earth, and the dust clouds and mud swirls had cleared, there it was: a cool, clear swimming hole, just like he'd promised us.

I WAS FIERCELY HUNGRY for reading material in those days. Bobby Strickland had a big steamer trunk full of comic books, and I dove into it headfirst. By the mid-'70s, Marvel Comics had a growing cast of Black superheroes. Black Panther was the first, followed by Falcon and Blade. There was even a Black girl superhero, Storm, the

"Weather Witch," who would grow up to become one of the first female members of the X-Men.

Comic-book superheroes reminded me of the Greek heroes, but instead of the gods tormenting them, they usually had some mad scientist messing with them and turning them all radioactive and mutant. What I noticed is how lonely they all were, how they had to keep their superpowers secret, because their secret powers were what made them mutants.

I knew from a young age that I was smart. I even suspected that I had special powers—though I couldn't quite say what they were. Not the kind of powers that could obliterate archenemies and ultra-bad villains. But I knew that being different that way—seeing things from unusual angles, having a hyperactive imagination, being able to count things without even thinking about it—was dangerous. It wasn't something I ever felt safe talking to anyone about, except Darin, who wasn't around anymore. If older dudes felt I was being a know-it-all or making fun of them for their lack of education, they could take it out on me with their fists.

The Black superhero who captured my heart was Luke Cage. He grew up in Harlem and got set up by a gangsta and sent to prison for a crime he never committed. To get out early, he agreed to be a subject in medical experiments in cell regeneration, which turned him into this freakishly strong guy who had "unbreakable skin"—so unbreakable that bullets would bounce off him. He became a "hero for hire." But he'd only work for good guys. I liked that Luke Cage was a good guy who could also be a tough guy. What really appealed to me was his invincible, unbreakable skin—which was his superpower. I made a note to look into that cell-regeneration trip.

By the time I worked my way through the whole trunkload of comics, I was ready for something grown-up.

15

DON'T KNOW HOW a copy of *Roots* ever found its way into Aunt Middy's house. I'd seen folks discussing the book on a news program on one of the two TV stations we got in Piney Woods. A white man interviewed a pale Black man wearing a suit—professor somebody-or-other from Jackson State—who said *Roots* was reclaiming the African in African American. Back in 1976, nobody was talking about African Americans. We were Black and Beautiful and proud of it, and ready to fight any cracker who called us nigger. But the African thing was new to me.

Whites and Blacks didn't live in the same neighborhoods back in the cities where I'd lived. In the cities, there was often tension and some hostility between the races. But there was none of the shuffling Negro bit that Black folks put on in Mississippi when they dealt with white people. Whenever the insurance man or the electricity man came around to get paid, or if someone had to talk to a store owner or just a stranger in town, it was "Yahssir" and "Yas'm," with head and eyes tilted down, or with a wide "you're safe with me, white person" grin taped across their faces. Everyone, it seemed, except my daddy, who was used to dealing with white folks back in New Orleans and looked them straight in the eye and talked to them exactly like he spoke to everyone else. As far as I was concerned, my daddy was an African American superhero.

Roots was a big-ass book, more than 650 pages with no pictures, and heavy in the hand like a Bible. I could tell the copy in Aunt Middy's house hadn't been read from the way the pages stuck together real tight, like the new social studies textbooks they unpacked and passed around the first day of fourth grade. Magically, no one was hassling me to do chores or trying to chase me out of the house the morning I found *Roots* wedged between the turntable and a stack of LPs. So I sat down on the living-room floor, right under the three framed pictures of JFK, MLK, and Jesus, and opened the book to the first page.

As soon as I began reading about Kunta Kinte, a boy growing up in a farming village in The Gambia way back in slave-time Africa, I crawled inside his story like it was a sleeping bag—and I zipped up that bag tight around me. Over the next few days I hid in every room of Aunt Middy's house so I could keep reading. When they flushed me out, I ran onto the porch or hid behind a tree, reading standing up so the red ants wouldn't get me.

The first thing that jumped out of the book at me was the scene where Kunta went into the forest to look for wood to make a drum and was captured by slavers. They knocked him out and he woke up naked and chained in a slave ship headed for America. Kunta's village of Juffure didn't seem all that different from Kelly Hill, and *I* walked through the woods alone all the time!

Reading my first real novel was like having a movie playing in my head. But it was richer than any movie or TV show I'd ever seen. I didn't miss having pictures in the book. I made up the pictures myself. But more than seeing, I could *feel* what Kunta felt when he was chained to a plank in the slave ship and trying not to poop and finally having to poop, and then having to lie in his own poop. I could smell it and I could feel his disgust and his shame. I could feel how hungry he must have been when he refused, because he was a Muslim, to eat the chunk of pork they gave him. And I felt his isolation when he was locked up by himself and so lonely that he captured a cricket just to have someone to talk to. And when Kunta set that cricket free, I felt like some locked-up part of me was set free too.

But my eyes didn't tear up until the scene when his owner took away the last thing Kunta still had from his African home to call his own: his name. That made me fighting mad. Being James Plummer Jr. made me feel special, and I didn't know what I'd do if anyone tried to take that away from me. Still, I wondered if Plummer was a slave name like the one his owner gave to Kunta.

I was glad I was in a private place out by the outhouse when I wept with anger for what they'd done to Kunta Kinte. I figured I'd never survive in Piney Woods if folks saw me crying. I was teased and bullied bad enough for being the kid who read books and made wisecracks and wore city clothes. I sure enough didn't want my daddy hearing from folks on Kelly Hill that James Plummer Jr. was a crybaby.

As soon as I finished *Roots,* I was crazy desperate to talk to someone about it. But I couldn't find anyone who'd actually read it. "I *woulda* read that *Roots* book," Uncle Henry told me, "but I heard they be making a television series outta it." As if only a fool would take the time to read a book if they could watch a TV movie with the same story.

I began to search for another big book to read. But books were hard to come by on Kelly Hill. Most folks just had the Bible and maybe a *Jet* magazine on the coffee table. Then, one weekend when we were over at Aunt Lilly's house for Sunday supper, I found the fattest book I'd ever seen. *The Rise and Fall of the Third Reich* was 1,248 pages! Aunt Lilly was using it to prop up her TV set, and I stared at the black swastika on its spine till it seemed to spin backward.

I had never heard of Hitler or the Third Reich. All I knew about World War II was what I'd learned from watching reruns of *Hogan's Heroes.* I read the "Rise" part in a week, totally hooked on the epic story of the Third Reich's world domination. Panzer divisions rolling into Vienna. All of Europe and North Africa collapsing before Hitler's swift brute force without barely putting up a fight. He was like some bad-assed gangsta with the audacity to hatch a scheme to take over the entire world, and the verbal pimp game to mobilize an entire nation to execute that plan.

Knowing nothing about history, I assumed Hitler was the hero of the story. Wasn't the main character always the hero? Only when I got to the "Fall" part did it hit me that Hitler was the bad guy, the supervillain who had killed millions in his quest for racial purity. I was embarrassed to be so ignorant.

The Rise and Fall of the Third Reich didn't make me cry. But it made me feel full, like I'd eaten all of Thanksgiving dinner all by myself. All that history, all those battles and generals and tanks and planes and submarines, were now living inside my brain. I'd always felt smart, but I realized my head had been empty. Now, with books, I could pour it full of information fast. Some days, I could actually feel my mind soaking up knowledge like a thirsty plant. As the blank parts of the world started to fill in for me—ships carrying slaves from Africa to the New World, battlefields across European and Asian countries I'd never heard of before—I felt less like a weirdo and more like a member of a secret sect of people who communicated across space and time in code.

That's when I realized you could be smart and ignorant at the same time. There were plenty of smart folks on Kelly Hill, including my parents and Aunt Middy. But because they'd never had formal schooling, because they seemed almost afraid of books, they didn't know about all the things going on and on and on in every direction, across time and around the world.

16

EVERY WEEKDAY MORNING, I'd walk across the cattle gap and down Kelly Hill with Andrea and Bridgette to catch the school bus. After bumping along rutted dirt roads for almost an hour with our heads knocking up against the ceiling, the bus would drop off Bridgette and Andrea at Quitman High School and, finally, me at Quitman Elementary.

Fourth grade turned out to be the year I put my hood on my back, as they say, and really motored through the books at school. We'd get a textbook in the afternoon, and that night I'd read till the end—then feel restless in class while it took everyone else weeks to catch up.

Unlike most of the schools in our area of Mississippi, Quitman Elementary was integrated. I didn't have any feelings one way or another about white kids. So I was truly shocked my first week of fourth grade, during a game of keep-away, when a kid yelled, "Keep the ball from the niggers!" No one had ever said that word in front of me in a schoolyard. At least, no white person had.

One day during recess, a group of us kids were playing on the monkey bars. I was making motor-revving sounds and shouting, "I'm Speed Racer!"—because Speed Racer was a cool Japanese anime series that we all watched on Saturday mornings.

"You can't be Speed Racer, because you're Black," said this red-headed white girl, who I had thought was cool. "Mark is Speed

Racer," she said, smiling over at a blond boy revving his engine on the far edge of the monkey bars.

I was thinking, That's fucked up. But what I said was, "All right, then, I'll be Racer X."

The racial split wasn't just on the playground. All the white kids sat together with their plastic-wrapped sandwiches in fancy lunch boxes. Boys had either Batman or Spider-Man that year, and the girls were deep into the Partridge Family. We Black kids sat together and ate our free lunch off of plastic trays.

In the classroom, the fourth graders at Quitman were divided into four tracks. All the white kids seemed to be in the first and second tracks, and all the Black kids were in the third and fourth tracks. Fuck that shit, I thought. They just put us here because they think they're better than us. I'll show them who's inferior.

To get from one level to the next, you had to complete a series of assignments and tests with zero mistakes. But if you made any mistakes, there were materials you could study before taking the test again. Since I loved to read, I was happy to hit the study materials as many times as I needed to advance levels. By the spring of fourth grade I'd worked my way up from the fourth to the first track and was the first to complete all the work. I felt pretty good about that, even though my white teacher put me down by saying, "James, your handwriting looks like chicken scratchin'." And just in case I was thinking of acting uppity, she pointed to the white kid who finished second and said, "Sarah was more careful and made fewer mistakes."

MAMA FELT BAD that she'd dropped out of high school at sixteen. So she started taking courses at Meridian Community College so she could take her high school equivalency exam.

One day I heard Mama talking to her friend Celia about how hard her physics class was. I started looking through her textbook, which was called *Physics Made Simple*. After working through a sample problem, I said out loud, "This *is* simple." It was logical in a way I understood in my bones: physics was just logic problems!

I had Mama sit down next to me, and I showed her how to solve the sample problem. But she didn't get it. Then I started doing other problems to show her more examples. She focused for a few minutes and tried, but she still didn't get it. "Come on, Mama!" I egged her on. "You don't see that? Look. Let me show you another one."

When she finally got it, I was all happy and proud of myself for being able to explain it to her. Mama held her head like it hurt and said, "Boy, you and yo' physics is giving me a headache. I'm gonna have to look at it again later." She watched *The Tonight Show* while I stayed there at the kitchen table working the problems till she told me it was time to go to bed.

WITH ALL THE READING and learning I was doing, I got a reputation as a know-it-all. But not so much in a bad way. Folks started calling me Professor. When older folks would get into arguments—usually about random shit like whether Mississippi was in North America or South America—they'd call me into the room to settle the debate. "Professor! What the books gotta say about it?"

It didn't matter what I told 'em. My answer would always bring out the same reply from the losing side: "Hell no! That boy don't know. I'm telling you . . ." Most of the time, I just kept my mouth shut so that an elder could save face. I'd learned that when I corrected a teacher in class, I'd get my ass kicked out of there, and quick! So when Uncle Henry said something like, "I hear tell dem hurricanes were cooked up by some voodoo spells cast down in Haiti," I just nodded and replied, "You don't say."

17

THE B&M CLUB was the coolest thing to happen around Piney Woods in a long, long time. Everyone said so. It was just a roadhouse before Daddy fixed it up into a proper juke joint. He hauled in new tables and chairs and replaced the felt on the pool table. He hung lights from the ceiling and walls, including black lights for the entrance and dance floor. He even built a DJ booth and hooked it up to four big speakers. Daddy knew how to do it all: electric, plumbing, tiling. I just handed him tools and helped clean up at the end of the day.

Like most juke joints, the B&M was technically a bring-your-own-bottle social club, which meant we didn't have a liquor license. In fact, we'd serve anything to anyone over twelve who could pay. Mama ran the bar, Andrea was the short-order cook, and Bridgette helped her make fish sandwiches, bus tables, and wash dishes.

The B&M was open Thursday through Sunday nights. I'd go with Mama and Bridgette and Andrea in the late afternoon and practice pool while they set up. Mama played blues, then switched over to R&B soul when folks started arriving. Men dressed up country fabulous with three-piece suits and bell-bottoms. The women wore their dresses short and tight, with big hoop earrings dangling below their Afros.

Mama switched off the lights—except the one over the pool table—and switched on the black lights, turning everyone's Afros into glowing balls. That's when I'd go to work hustling up some pocket money at the pool table. I'd challenge someone to play me for a dollar a game. Usually they couldn't resist what looked like easy money.

I set them up by dishing out some outrageous raps inspired by Rudy Ray Moore records that were making the rounds in those days:

> First I'ma let you break, and pray that you sink one or two,
> 'Cause you ain't got no damn idea what I'm 'bout to do to you;
> I'm so muthafuckin good I don't need no chalk or stick,
> I can just run all the balls using the head of my big hard dick!

Then I'd break hard, drop a ball in the side pocket with the lightest kiss of the rock, and carom a bank shot into the corner. I could even shoot the cue behind my back, though I was too short to keep both feet on the floor. After dropping a few bucks, folks would get annoyed at the trash-talking nine-year-old and his hustler's trick shots, and they'd move me off the table.

By nine or ten at night, when the club was filling up with bodies and smoke and '70s soul, Mama would send me up the road to Aunt Lilly's house to sleep.

I WAS ASLEEP at Aunt Lilly's the night the bad stuff went down, so I only heard about it afterward. Most Saturday nights, Mama and Bridgette would pick me up on their way home after closing so Bridgette and I could go to church the next morning at East Baptist. I knew something was messed up when I woke up at Aunt Lilly's on Sunday morning, and she was making me breakfast.

By the time she drove me home later that morning, Mama and Bridgette were busy packing up our stuff. Mama paused just long enough to explain to Aunt Lilly that Daddy didn't want her in Clarke County when the sheriff came around to question folks who were at the club the night before. Especially not someone who'd been tending bar in a dry county and in a juke joint that didn't have any sort of license to operate.

Turns out that sometime after midnight there was a ruckus at the B&M Club. The usual thing, somebody hitting on some woman who wasn't his to hit on. Followed by the usual threats and trash-talking. A knife came out, and then a gun. The dude with the gun shot the dude with the knife square in the chest. He fell over holding his chest and bleeding in the middle of the B&M's new dance floor. Before Dazz's "Brick" had finished playing, the dude with the knife was stone-cold bled-out dead.

Everyone split before the cops showed up. Daddy arrived in time to pack up all the booze in his pickup. But he had to leave the stereo equipment behind, along with all the black lights and such.

So that was the end of the B&M Club and the end of living on Aunt Middy's farm. We didn't even have time to say goodbye to folks before Mama and Bridgette and I scooted out of Piney Woods and back to New Orleans.

18

FANTASIZED THAT since we had to move back to the city just as summer was coming on, we'd land in an air-conditioned apartment complex with a pool in New Orleans East. But that was just my overactive imagination at work. All we could afford was a tiny two-bedroom flat clear across town in the Ninth Ward, right on the edge of the Desire Projects. Better known as "The Dirty D," the Desire Projects were totally cut off from the rest of New Orleans, pressed hard between the railroad tracks, the Mississippi River, and the Industrial Canal. Nothing green grew there. No trees or grass or shrubs. Just thousands of down-and-outers like us—mostly Black, and dirt-poor.

You didn't go wandering around the Desire Projects. Not in the daylight, and certainly not after dark. "That's where you go to get killed," was the standard warning from kids at school. Whenever I took my skateboard outside to practice my tick-tack techniques, I made sure to steer clear of the junkies and gangbangers on the sidewalks. I waved at the girly guys who called out low and sweet, "Hey, Li'l Jame. Skate over here, pretty boy," but never slowed down or stopped like they asked me to.

Daddy put an old window unit in Mama's bedroom the day after we moved in, but the front room had only a small fan that sat on a folding chair next to the couch. If the window unit broke down and the apartment was sweltering hot, Mama sat on the couch in front of the fan, holding a small face towel in one hand to wipe her sweaty face, and a Kool in the other.

Mama felt like a prisoner in the Dirty D, with no work and no friends nearby. It got to where she started talking back at the TV, sassing old Walter Cronkite while he read the nightly news. "What damn economic recovery, you ol' fool? Ain't no muthafuckin' jobs. And prices is rising so fast I can barely pay the 'lectric bill." I just tried to stay out of Mama's way when she got like that.

The worst part of living in the Ninth Ward was being so far away from Darin. He was all the way across town in the Goose. Even if I could get a ride to the Goose, Darin was involved in football and other organized sports by that time, so he was never home after school or on weekends.

Even Bridgette ditched me that summer. She'd fallen hard for a guy from Mississippi named Dwane Morgan, whom she met in the B&M Club that last night when the dude got shot and killed. I heard her tell her friends they'd been dancing close together when the shots rang out and Dwane had protected her and hustled her outta there. Ever since then, Dwane had been calling her on the phone and driving all the way to New Orleans to see her.

Now that she was sixteen and had a steady boyfriend, Bridgette didn't seem to have time for a ten-year-old little brother. When she wasn't taking a city bus across town to visit friends, she'd hang out in Mama's bedroom daydreaming about Dwane or talking on the phone. I wasn't allowed in Mama's room, ever. When I tried to talk to Bridgette through the door, she'd open it just wide enough to shoo me away, just like Mama did.

I was as lonely as I'd ever been, and bored as hell. There wasn't a single book to read in our apartment. What saved me was a door-to-door salesman who came knocking one hot day in August. Talk about angels in disguise. This guy wore the same short-sleeved shirt and clip-on tie as any other salesman selling deluxe leather-bound Bibles or newfangled vacuum cleaners with space-age names. Ordinarily, Mama would never unlatch the safety chain for "one of dem smooth-talkin' con men." But when this one passed her a glossy full-color brochure through the crack in the door, she let him in.

You see, Mama was always trying to get some respect from her older sister, Aunt Jean. Back when we'd stayed with her awhile in L.A.—before she and Mama got into a brawl about something and we had to leave in a hurry—Aunt Jean would brag about having a complete World Book Encyclopedia set. "That's what makes a house a home," Aunt Jean would say in her stuck-up voice. I was too young to read in those days back in L.A., but I remember spending a long afternoon at Aunt Jean's house staring at pictures of snakes in the "S" volume.

So when the salesman showed up that day with his friendly grin and a slick World Book brochure, Mama looked around our apartment with its saggy old couch and Salvation Army table—and then she fished a sweaty $5 bill from her purse for the first installment.

A week later the whole set arrived packed in five cardboard boxes: twenty-two white leather-bound volumes lettered A to Z, with J–K, N–O, Q–R, U–V, and W–X–Y–Z combined into single volumes. Since we didn't have a bookshelf, I lined the World Books up alphabetically on the floor with a brick on either end to hold them up. I had to agree with Aunt Jean: they did make it feel like a home.

I couldn't wait to dive in. Since I didn't want to miss anything good, I decided to start at "A" and read every volume in alphabetical order. After reading the "Aardvark" article, I wouldn't shut up about the burrowing and feeding habits of the only living species of the order *Tubulidentata*. When Mama and Bridgette said I was driving them crazy with all my chattering about aardvarks, I moved on to albatrosses and anteaters. Even though she made a sour face every time I looked up from a volume to say, "Mama, listen to this," I could tell she was proud to finally have a World Book set in her home—and someone who was actually reading it.

19

Every once in a while, Daddy would visit us in the Ninth Ward. But there weren't many laughs between Mama and Daddy, and there were constant arguments about money. Daddy had helped her pay to move into the apartment. But after that he didn't give her much more than the cash in his pocket during his visits, which became fewer and further between as the summer wore on.

Daddy usually showed up with a small cooler of beer and immediately started rolling a joint at the kitchen bar. Things lightened up for a bit when Mama drank the ice-cold beer and they shared the joint. I'd happily pop to my feet and carry the joint from Daddy at the bar to Mama on the couch, and back again.

Soon enough, though, Mama's mood would turn sour and I'd get kicked out of the apartment so the grown folks could "talk"—which meant they were most likely going to argue. I wasn't allowed to return until I saw Daddy leave. That was a problem if he arrived near sunset, since I didn't want to be on the sidewalk after dark.

My safe spot outside the apartment was the stairwell of our building, between the first and ground floors. That cubbyhole became my World Book reading room, my go-to escape hatch from everything nasty and scary in the Ninth Ward. Armed with a leather-bound volume and a flashlight—because the stairway bulb was always broken—I could spend a whole afternoon or evening going down whatever path the World Book led me.

THE DAY THE BIG THING HAPPENED, my fifth grade elected our class officers. Since I was the teacher's pet, Miss Hall piped up, "Doesn't anyone want to nominate the best student in our class?" until someone finally did. I was super excited and already dreaming about being Class President Plummer with everyone having to call me President. Then they counted the ballots, out loud, in front of the whole class. I came in last among the three candidates. I got one vote. My own. That's how popular being the class brain made me in the Ninth Ward.

I was crushed to discover that the only person who liked me was Miss Hall. I couldn't get out of school fast enough after the final bell. Pressing past the crowd of sweaty bodies in the hot, humid hallways, I ran nonstop for three blocks till I was sure I wouldn't have to see or talk to anyone I knew.

By the time I got home, I was dripping with sweat. I just wanted to cool off and shut down my mind. So I stripped to my shorts and sat down in front of the fan with a mason jar of ice water pressed between my legs. I tried to concentrate on the numbing cold on the inside of my thighs and block out every other feeling.

Just then, Daddy and a friend—a white guy I'd never met—came through the front door. Daddy didn't introduce me to his friend or ask me to tell him my name out loud. He shouted "Laney!" through the bedroom door. Mama came out, still tying her hair in a knot on top.

Mama kicked me out of the apartment straight off. I pulled my pants on, grabbed the "E" volume of the World Book and a flashlight, and scrambled out and down the stairs. I scanned the landing with my flashlight to make sure there were no roaches or rats there. Then I settled in to read. The air was stifling and hot, but the concrete steps felt cool against the backs of my legs. Curled up inside my little funnel of light, I opened the "E" volume to where I'd left off. I read through the whole article about eggs—including the subsections on bird eggs, reptile eggs, monotreme eggs, and "eggs as food." It was pretty fascinating stuff. And it took my mind off the class election.

The next full-length article was about Albert Einstein. Ordinarily, I'd speed-read through a biographical entry about some white dude born in Germany a hundred years ago. But right away I felt connected to Einstein. I could tell from his photo with that wild hair that he was weird, like me. And probably a loner.

According to the article, Albert was so slow to learn to talk that as a kid they called him "der Depperte"—which is German for "Dopey One." Not as bad, I thought, as my mama's nickname for me: "Shit-ass." Even though some folks told me I was book-smart, they were usually the same ones shouting at me to "Wake up!" and "Stop staring into space!" and "Quit counting everything and saying every weird thing that pop into yo' head!" The article also said that Einstein used to get in trouble for sassing his teachers, just like I did. And his family moved a lot when he was a kid, just like mine had.

Albert Einstein and I probably would have been friends, I thought.

Then the article introduced relativity, and Einstein's famous equation $E = mc^2$. Relativity was nothing like the physics I'd learned in *Physics Made Simple,* my mother's textbook back in Piney Woods. It was much wilder. When he came up with $E = mc^2$—energy equals matter times the speed of light squared—Einstein revealed that mass *is* energy. And that energy can have weight and create gravity just like mass.

Einstein realized that space and time were not completely separate, and he combined them into a four-dimensional "spacetime." Everything in the universe is moving at the speed of light at all times. Not just through space, but through space and time: which is spacetime. If you move faster through space you must go more slowly through time. This means that there are no true distances or time durations. *All distance and time is relative.*

At the moment he realized time and space were relative, Einstein said, "A storm broke loose in my mind." And that's exactly what happened to me! Right there in that hot, dark stairwell, I had my own brainstorm. I'd always known that time could feel like it was moving slower—like in a boring class—or faster, like when Darin and I were playing touch football and an hour went by in a few minutes. Einstein discovered that the actual passage of time could change. He called it "time dilation," which means that the faster you travel through space, the slower you travel through time.

But it wasn't just time that was relative. In spacetime, time *and* space could bend, contract, and stretch. I knew it!

Part of me had always known that *things were not as they appeared*. An alternate reality existed just next door, just out of sight and just out of reach. I wondered if relativity explained why even though my body was stuck in a crappy apartment across from the Dirty D, my mind could carry me a million miles away from the Ninth Ward, the hoodlums on the corner, Mama's sadness and anger, my own loneliness.

The end of the World Book article said, "For more, see article on 'Relativity' in Volume 'Q–R.'"

My mind was racing so fast I could hear it whir in my head. I clicked off the flashlight, and sat in the dark a moment to try to slow things down. But even in the dark my thoughts were moving at the speed of light. I clicked the flashlight on and off to see if I could trace the light beam as it traveled to the far wall of the stairwell and bounced back to my eyes. Of course I couldn't. But in a nanosecond my mind traveled upstairs to where I could see the "Q–R" volume, nestled in between volumes "P" and "S." I jumped back into my body and bounded up the stairwell to retrieve it.

20

BURST INTO THE APARTMENT to find the white guy sitting on the couch cracking open a can of beer. I slipped past him, slotted the "E" volume back in its place, and grabbed the "Q–R" volume.

I could hear Mama yelling at Daddy in her bedroom, "I can't take this shit no mo', James! Something gotta change. I helped you run yo' club, now you gotta help me, goddammit. Bills don't pay they damned self."

I peeped through the half-open bedroom door. "C'mon, Laney," Daddy said. "Everythang gon' be okay. You just need to settle down."

When Mama saw me through the open door, she shouted, "Shit-ass! Didn't I tell you to stay outside?" She started toward me, and Daddy grabbed her by her arm. "Let me go!" she growled, turning to face him. "*You* takin' him over Christmas *and* next summer when school get out," she said, her voice low and angry.

"Sorry, Mama," I said, and scooted out the apartment door.

BACK IN THE STAIRWELL, I flipped through the "Q–R" volume till I got to "Relativity." I could feel my heart thumping in my chest as I traced the flashlight beam across the page, my brain tripping out on the magical stream of photons flooding out of the end of my flash-light and lighting up the encyclopedia article.

Einstein figured out that light moves faster than anything else in the universe *because it has no mass*. So that's what makes light so damn light on its feet!

$E = mc^2$ means that what we call mass is not just material "stuff." Most of the weight of matter is energy that's locked up inside the cores of atoms, in the forces that bind them together. When you free up all that energy locked up inside an atomic nucleus . . . BOOM! It all comes rushing out as light and heat. That's where all the energy coming out of nuclear bombs, the sun, and the stars comes from.

Relativity explains that energy escaping from mass is powering the entire universe—warming the Earth and growing the crops on Aunt Middy's farm. And at the same time, the curvature of space-time is moving the tides and determining how fast we grow up and grow old.

No way! I thought. Is this real? I'd always been hooked on the magical feats and weird happenings in the Bible and in Greek my-thology and comic books. But I never believed they were real. I couldn't turn water into wine, and neither could anyone else. I couldn't fly, like Iron Man, or shoot spider webs from my hands, or turn invisible, no matter how badly I sometimes wanted to. And I certainly didn't have unbreakable skin.

Einstein's superpowers were all in his mind. And right then I was holding the superpower of relativity inside my mind too. I felt like Superman when he first realized, as the boy Clark Kent, that he'd acquired special powers when he arrived on Earth from Krypton. Even though I was a poor, uneducated kid from Smallville, I could lift cars with my mind!

If relativity was science, not science fiction, then I should be able to observe it.

"I need my skateboard," I said out loud.

My flashlight beam caught a roach scurrying across the stair be-neath me. I quickly stood up and crushed it with my foot. Its exo-skeleton popped with a crunch and squish, which sent a chill down the back of my neck.

Just then, I heard our apartment door open. "Let's go, man," the white guy said. "She's gonna be all right."

I stood up and pressed my back against the wall as Daddy and his friend came down the stairs. "Man, she crazy," Daddy said to the white guy. "Why'd she have to go do that?"

"I looked at her hand," the white guy said. "It ain't broken."

They walked past me as if they didn't even see me. I looked at the stair where the dead roach was lying and hoped they wouldn't step on it. Daddy did. It stuck to the bottom of his shoe and then detached in two pieces on the stair below.

I opened the apartment door to find Mama sitting on the couch with her right hand stuck in a bowl of ice water. I could see through the drywall where she'd punched a hole in it. I felt sorry for Mama. But I didn't want to know about that hole. All I knew was, when things got hectic with Mama, they usually got hectic with me. I just wanted to grab my skateboard and get out the door.

"Get me a cigarette, baby?" Mama asked. When I picked up her pack of Kools off the table, she lifted her bruised hand halfway out of the ice water. "Light it for me, will ya?"

I pulled one out of the pack with my lips. That's when I noticed everything seemed to be moving in slow motion. I tore off a paper match and dragged its red sulfur head across the striking strip. The spark roared into flame in super slo-mo. Bond-energy freed! I touched the flame to the end of the cigarette till it crackled and glowed—more bond-energy freed! I sucked in a mouthful of mentholated smoke and blew out a plume that hung in the air like a sweet-smelling question: *Is this what spacetime feels like?*

I tucked the filter end between Mama's lips, kissed her on the cheek, and beat it out the door with my skateboard.

21

WHEN I GOT DOWNSTAIRS to the sidewalk, the corner street-light had just clicked on. That was usually my signal to get inside. But everything felt different now. Energy seemed to be bursting through every crack in spacetime. Not just from the street-light and the passing headlights. Every car and living thing—even the skinny old alley cat slinking around behind the trash cans—seemed to be crackling with energy.

The usual assortment of badasses were spread out lengthways across stoops, leaning against lampposts and parked cars with their legs blocking half the sidewalk. But I wasn't afraid of them. I stepped over them on the front stoop, slapped my board on the pavement, and started to glide.

"C'mere, little dude," one of them called to me from the side of a parked car. I just rolled right by and didn't pay him no mind. I was on a mission.

As I approached the corner, I heard a familiar refrain.

"Heyyy, Li'l Jaaame!"

I'd usually just wave as I passed the girly guys with their scarves tied around their heads in a bow in front and their eyebrows painted up high on their foreheads. Not tonight.

"Y'all wanna help me?" I asked her and her two girlfriends.

"Help you with what, li'l badass boy?"

"What you gon' do for us if we help y'all?" asked her friend.

"What you need, baby?" asked the third.

"I want to slow down time," I told them.

"Well, hell," said the first one, swiveling her neck and planting her hand on her hip, "we *all* wanna do that! How you s'pose you gon' make that happen?"

"With my skateboard. All I have to do is skate really fast and then throw a rock straight up, then catch it. And all y'all gotta do is stand where I tell you to and count how long the rock stays in the air."

"C'mon, y'all," she said. "Let's help this boy build his time machine."

"Not a time machine," I said. "Spacetime."

"Pfft! Boy, please! Spacetime my ass. I ain't got no time for this shit."

"Shut up, dumbass," another one said. "This boy serious, y'all. They say he a smart one. And look at those eyes on him."

"Oh. I am lookin'! Them lips too."

"I told you leave that boy alone, nah! Gon' 'head, baby. You was talking about spacelight."

"Not spacelight," I explained. "Space-*time*. If you move fast enough through space, it makes you go slower through time. That's how the universe works."

"Okaaaay . . . what you need us to do?"

"Watch this," I said. "I'ma throw this rock straight up and then catch it. If I throw it the same way to the same height, it'll always be in the air for the same amount of time, no matter how fast I skate. I'll throw; y'all count."

"I got a better way!" one of them said excitedly. "We can sing 'Crawdad Hole.' Start when he throw the rock up and stop when he catch it. The word we on in the song at that point is how long the rock was in the air."

I counted them off, "One . . . two . . . three. . . ." Then I threw the rock up in the air as they began singing: *"You get a line and I'll . . ."*

On the word "I'll," I caught the rock and they stopped singing.

"Yes!" I said. "Just like that. Now, we gotta do it a few mo' times to make sure it's always the same."

I threw the rock up and they sang three more times. Every time, the rock fell in my hand on the same exact word in the song.

"All right, that's regular time," I said. "Now we gotta show how time is longer when I'm moving fast."

"Ooh, we wanna see you move fast, Li'l Jame! Can we chase you?"

"No," I said. "You each gonna have a different job in this experiment."

I placed one girl at the spot on the sidewalk where I'd throw the rock up as I passed by on my skateboard. The second girl's job was to stand next to the first and start singing when I skated past. I asked the third girl to run over and mark the spot where I caught the rock.

After I lined them up, I got on my board and skated down Scott Street as fast as I could. When I got to the first girl I threw the rock up in the air and they began singing "Crawdad Hole." When I caught the rock the third girl ran over to mark the spot. I did this three more times, skating at different speeds. Each time I caught the rock at a different spot, but at the same point in the song.

"That's it. We did it!" I yelled after the fourth time.

"If you say so, Li'l Jame," one of them said. "If you ask me, that's just keepin' rhythm."

"No, no. We just proved there's spacetime," I said excitedly. "When I was moving on my skateboard, I caught the rock in a different place than when I threw it up. So even though it was in the air the same amount of time, it must have traveled farther than when I was standing still. That's what we showed."

"Mmm . . . hmmm. When you talk like that, Li'l Jame, you get all worked up."

"Since I was traveling with the rock," I explained, "I saw it go straight up and straight down, which is a shorter distance than what y'all saw. The time y'all measure between me throwing it up and catching is longer than the time I measure. We slowed down time!"

"I don't know about slowin' down time, but we done killed some time, that's for sure," one of them hooted.

I skated home as fast as the super slo-mo of spacetime would let me. When the dudes out front saw me coming, they actually made way for me, like I was a force of nature they couldn't stand up to.

I ran all the way up the stairs and into our apartment, bursting to tell Mama all about my thought experiment.

Mama looked up from behind a half-filled suitcase. "Pack up yo' things," she said. "We movin'."

22

W E LANDED IN a ramshackle old shotgun house on Dale Street in New Orleans East. It belonged to Mama's cousin, who'd inherited it from his grandfather. But it was in such a run-down condition no one wanted to live there. Mama's cousin said we could stay for free if we cleaned it up.

The rear room of that shotgun house on Dale Street was the first time I had my own bedroom. It was small and crawling with mice, giant spiders, and weird lizards, but a perfect place to finish reading my World Book encyclopedias. I powered through volume after volume, soaking up the worlds inside them. One night I looked up from the "N–O" volume, where I'd been reading about Isaac Newton and his deterministic physics, when I realized the sun was coming up.

After I almost set the house on fire mixing together household cleaning supplies to investigate bond-energy release, Mama got me a chemistry set for my eleventh birthday. So my library bedroom became a laboratory bedroom, where I started conducting real experiments with little beakers and flasks. For a few weeks, I was enjoying life on Dale Street. Then, all of a sudden, the fun got up and left.

I could tell something was wrong because Bridgette had been acting strange all week. One night we were all in Mama's room watching *The Jeffersons*. Mama and I were laughing, but Bridgette mostly sat quietly staring at the floor like she was thinking hard about something. Then during a commercial, she told Mama they needed to talk, and she told me to go outside. I said I was going back to my room to read, but I stood just outside the kitchen door and leaned in to try to listen.

I could hear Mama yelling at Bridgette. Usually, I was the one she yelled at. I went back inside to find Bridgette and Mama both crying side by side. Mama rocked back and forth on the edge of the bed wailing, "Lord, Lord . . . my mama got pregnant at sixteen . . . I got pregnant at sixteen, and now you pregnant at sixteen. Lord, when are we ever gonna break this cycle?"

Mama waved me over. "Come here, baby." And for the first time I could remember, my mother pulled me into her bosom.

Bridgette held her head in her hands as she sobbed quietly, hiding her face. It felt weird. Mama was always switching up extreme emotions, but Bridgette never cried. I didn't know what to make of Mama's hugging me either. All that crying and hugging made me nervous. What did this all mean, and how was it going to play out for *me*?

One day when Mama wasn't home, I listened in as Bridgette told her friend Toni how she got pregnant. "It was my first time. Dwane had come to see me and took me out just like normal. But then instead of taking me home he drove me to his cousin's house, saying no one was home but he had to go in there for something. Then he opened my door, picked me up, and carried me through the front door and into the house—like I was a bride. He carried me all the way into the bedroom and laid me down on the bed."

"What did you do?" Toni asked.

"I didn't do nothing. I just let it happen. Then I missed my period. It was just that one time."

"Do you love him?"

"I do."

"Are you gonna have the baby?"

"I am."

"Why?"

"I ain't gon' kill my baby. I'm having it. He said he gonna marry me."

"Girl, you sure? You only sixteen. If you don't have this baby you can finish high school and get a good job. You already know how to type fast."

"I got to get out of here. I'm tired of Mama. I'm tired of her men. I'm tired of moving every few months. I'm ready to settle down, and if that means getting married and having this baby, then that's what I'm gonna do. I'm leaving and I'm not comin' back."

Hearing that broke my heart. What about me? I thought. I understood her wanting to get away from Mama, but why would she want to leave *me*? If Bridgette left, I was going to be alone with Mama. Who was gonna feed me and get me up in the morning for school? She's just saying she leaving, I thought. I hoped. Besides, no way Mama was going to let her get married. But I was wrong.

Mama didn't want Bridgette to have the baby. But Bridgette's mind was made up. From then on, Mama was a nervous wreck. She barely ate and just seemed to smoke Kools all day long. While Mama got skinny and depressed, Bridgette got round and happy. She was at the house less and less.

Then one day that spring, with a big smile on her face and a baby swell in her belly, Bridgette stepped out of Dwane's trailer wearing a long white dress and a white lace cap on her head. She exchanged marriage vows with Dwane right there in the yard behind his trailer while a few dozen family members stood and clapped. When it was over, she got in the passenger seat of Dwane's freshly cleaned and waxed car. He got in the driver's seat, smiled at Bridgette, and they pulled away. She didn't even say goodbye to me. She just waved over her shoulder and drove away.

23

THINGS GOT DIRE after Bridgette left to live with Dwane near Heidelberg, Mississippi. I couldn't take care of Mama, and she couldn't take care of me. She'd leave $2 in the kitchen with a note telling me to buy myself a can of pork 'n' beans for dinner—then disappear for the night, and sometimes two nights.

So I was relieved when Mama told me she needed a break from being a mama and was sending me over to Daddy's place to stay as soon as school let out for summer. I'd only slept over a few nights at his house since we got back to New Orleans, and now I was going to get six whole weeks with him!

During the days, I was pretty much on my own. Andrea, who had graduated from Quitman High by then, was living with Daddy and working at a beauty shop. Daddy was still working weekdays at the Kaiser plant, and his new young wife, Stephanie, worked split shifts at the hospital.

Daddy's house in a middle-class neighborhood called Little Woods was nothing like our shotgun shack on Dale Street. His place had air-conditioning, four bedrooms, a two-car garage, a living room, and a den. There was always food in the fridge—even breakfast food. There was also a whole lotta weed and guns around the place.

Andrea explained Daddy's New Orleans weed business to me. It wasn't anything like his Mississippi grow. He imported weed in bulk from Colombia, Mexico, and Thailand, and kept big bales of it in the garage wrapped in plastic. More big bags of it were stashed under my bed and in my closet. The Colombian and Mexican weed was pressed into hard bricks, while the Thai came wrapped tight around sticks that were packed into shoe boxes.

My job that summer was cleaning the weed—sort of like at Aunt Middy's, but more of a two-fisted style of weed cleaning on account of how hard-packed it was. There were three big pink plastic tubs on the floor of my room, each filled with a different kind of weed. But Daddy didn't mess with nickel and dime bags. He only sold it by the ounce, quarter pound, or pound. His clients were other dealers and guys who wore suits. Businessmen and politicians, he said. He even had white customers.

Daddy explained to me that he was a living-room dealer, not a street dealer—and he schooled me in the difference. "You never wanna dress like a drug dealer, act like a drug dealer, or hang out where drug dealers hang out. Don't ever be a street dealer," he'd say. "Be a living-room businessman. Treat everybody with respect and never cheat nobody."

Anytime someone came over to score, Daddy took precautions before letting them in. "If you don't know 'em," he explained to me, "dey either police or dey there to rob you—at least until you find out otherwise."

When there was a knock at the door, Daddy would fetch three pistols from his bedroom. He'd put one, a big-ass .44 Magnum revolver, on the hall shelf next to his autographed Saints football. Halfway down the front hallway, he hid another pistol behind a plant on top of a sideboard. The third one he tucked inside his waistband at the small of his back. I was used to seeing Daddy handle guns in Mississippi. "If you don't hunt it, you can't eat it" was one of his sayings. And Aunt Middy always kept a pistol under her mattress and a shotgun in the closet. But these weren't country-style guns. They were the mean and nasty types you saw in gangsta movies.

Once he verified it wasn't the police or someone worse at the door, Daddy switched over to his normal friendly self. But still care-ful. "Make sure you always have eyes on 'em when dey in your house. Even if dey ain't the police or trying to gaffle you for your dope, they could have some other angle. Dey got all kinda games in dis town."

Once Daddy was done with his business, he liked to call me into the living room. "Tell 'em yo' name, boy."

"James Plummer Jr.," I'd say proudly. They'd laugh along with Daddy when I said that and dap each other up. But they never told me *their* names.

It FELT GOOD to be living with Daddy and Andrea in a real house, and to be a help to Daddy with his living-room business. Also, Little Woods was near Darin in the Goose, so I even got to see him once in a while.

Halfway through the summer, Daddy dropped me off at the Browns' house for one of Miss Jeannie's fried chicken dinners with red beans and rice. Darin and I competed to see who could eat the most, and after dinner we all got a game of bid whist going. It must have been almost midnight when the phone rang.

Miss Jeannie answered, and I could tell right away that some-thing bad had happened. She just listened and didn't say anything but "Lord, no!" and "All right, I'll keep him here for the night."

She didn't want to tell me what had happened, but I kept on her till she did. "Somebody robbed yo' Daddy's house tonight, and they shot yo' sister." She choked up and lost her words in the tears that streamed down her face. "James is at the hospital with her now. She going to live, they say."

Daddy picked me up the next day and drove me to Mama's house on Dale Street. We drove in silence for a long time. I asked him what happened to Andrea, and he didn't say anything for a while.

When Daddy finally spoke, it was in a quiet voice I'd never heard out of him. He told me someone came to the door and Andrea, who was alone at home, let him in. They pulled a gun on her, then backed a truck into the garage and loaded up all the stuff. One of them asked Andrea where the money was, and when she didn't say nothing, he shot her in the stomach. The doctors stopped the bleeding and she was going to be able to leave the hospital in a week or two. But because of where the bullet hit her, she wouldn't be able to have babies.

"She described 'em. I know who dey is," Daddy said. "After folks see what happens to dem boys, ain't none a their kind gon' ever dare step foot in Little Woods again."

He drove for a while without saying anything. "Some folks, dey animals. Dey give you dap with one hand and stab you with the other." Daddy shook his head side to side and for a minute I thought he might cry, which made me anxious. So I looked out the side window and counted the telephone poles all the way back to Dale Street.

24

McDONOGH #40 ELEMENTARY SCHOOL was two blocks from our house on Dale Street. After the principal walked me to my classroom, I overheard him tell the teacher, "Keep your eye on this one. His mama says he's really smart and good in school." It felt good being introduced like that, but I worried my classmates would put a hate on me for knowing the answers or getting special attention from the teacher.

My sixth-grade teacher at McDonogh 40 took a liking to me right off. Especially after I did a six-page report on the human brain, complete with multicolor diagrams. First, he asked me where I'd copied my report from. When I told him I didn't copy it from anywhere, that I'd read six different articles in the encyclopedia to learn it, he said he believed me.

The next day after class he told me he'd recommended that the school give me an IQ test, and the first part was with a psychologist. He led me to a room where a white man in a dark-blue suit was waiting for me at a table.

"Hello, James. How are you today?"

"I'm good."

"I'm going to ask you some questions and then you'll solve some puzzles and play some games. How does that sound?"

"Fine," I said.

He handed me a manila folder and said, "Hold this up in front of your eyes and I'll ask you some questions." I saw that he had a sheet of paper with the questions and answers already typed on them. He read the first set of questions. "Tell me what kind of book the following scientists would carry: A zoologist."

"A book on animals."

"A botanist."

"A book on plants."

"A paleontologist."

"A book on dinosaurs?"

"Good," he said.

"An ornithologist."

I'd never heard of that. But then I remembered that the sixth labor of Hercules was to kill the vicious Stymphalian Birds, who the Greeks called Stymfalides Ornithes. So I guessed: "A book on birds?"

"Wow! Really good."

We continued like this. Next, I completed a written multiple-choice test and worked a few puzzles that were increasingly difficult. I wasn't sure how I'd done. The following day I went back to the same room and was met by another psychologist, who gave me an even tougher exam. I took the test, then forgot about it. But a week later, when I got home from school Mama came running up to me and hugged me around the neck with both arms.

"My baby's a genius!" she gushed. She took me up into her bosom and started swinging me around like we were dancing or something. "I knew it," she said. "I always told everybody you was a genius. My baby's IQ is 162! Wowee!" Mama grinned from ear to ear. Then she balled her hands into fists in front of her and did a happy dance. "My baby! Shit-ass, I knew you was special. One day the whole world gon' know it."

I looked at Mama like she was actin' crazy, but I made sure not to come off disrespectful. Mama was happy right then, and I wanted to keep her that way.

WHEN I GOT TO SCHOOL the next day my teacher told me that because of my high IQ, I was going to leave class a couple days a week and take part in a special program for mentally gifted students. Now that I was certified as mentally gifted, I imagined things would go better for me.

But nothing ever seemed to stay good for long. It's like Newton said in his third law of motion: every action has an equal and opposite reaction. The universe needs to stay in balance. It wasn't long before ol' Newton came looking for my debt to be paid.

I had just started the second month of school when I came home to find Mama on the phone with one of her sisters in L.A. "I don' wanna send him to y'all, but I don't know what else to do. Bridgette probably gon' lose the baby if I don't go tend to her. I can't ask Dwane to take care of Bridgette, me, and him. . . ." When Mama saw me in the room, she turned to face the other way. "You sho' you don't mind? Yeah, I got train fare for him."

When Mama got done with her call, I could see she was crying. But I didn't want to comfort her. She was going to send me away.

"Look," she said, lighting a Kool and staring out the window instead of at me. "Your sister is about to lose her baby. She has to stay in the bed all the time. I need to go to Mississippi to take care of her. So I gotta send you to go live with Aunt Jean in L.A."

"By myself?"

"No. Not by yourself. Aunt Jean will be there. Your cousin Cheryl will be there. And you'll have your friends at school. I know you know how to make new friends. You do it every year."

I didn't know what to say, and I was too old to cry. It was the first time we'd had our own place to live, and I finally had a special program at school. Now she was sending me back to L.A., where all I remembered was beatings and gangs and family that fought with each other like gangstas. And she was sending me by myself to stay with an aunt who only ever disrespected us. I was always looking to please Mama, and not be the angry one in the house. But I felt good and angry right then.

I started to make a list in my head about what I'd be able to take with me to California. Usually, we'd have the whole Maverick to fill with stuff, but I was going on a train, so I'd have to fit everything into one or two bags. There definitely wouldn't be room for my chemistry set, or the encyclopedia.

I went back to my room and found a book of poems I'd been reading by Edgar Allan Poe—another weirdo I felt a kinship with. In case I didn't have room for the book in my suitcase, I started memorizing one of the poems that spoke to me . . .

"Alone"

From childhood's hour I have not been
As others were—I have not seen
As others saw—I could not bring
My passions from a common spring—
From the same source I have not taken
My sorrow—I could not awaken
My heart to joy at the same tone—
And all I lov'd—I lov'd alone—

I put myself to sleep that night by reciting it to myself. And the whole train ride to L.A., I heard it playing back in my head.

PART TWO
COMING OF AGE IN MISSISSIPPI

Where I'm at, if you're soft
    you're lost,
Cuz to stay on course means
    to roll with force.

—KRS-One

25

A YEAR AND A HALF after Mama sent me out west, I returned to New Orleans a different person. I was only twelve, but now I had body hair. I was bigger, meaner, and harder. In short, I'd gone bad. At least that's what folks said about me when I rolled by them on the streets of L.A. and Houston.

I had to go bad just to survive. During my sixteen months out west, I lived in nine different households and attended five different schools across California and Texas. Some of the folks I stayed with were relatives, and some were friends of Mama's or so-called friends of the family. None of them treated me like family.

At one house, my reputation as a smart kid preceded me. The father and his teenage son also had reputations for being smart and wanted to show me up, so they each challenged me to a chess match. I beat both of them in under fifteen moves. After that, I was sunk. The father turned me into the family's Cinderfella, making me wash dishes and clean the kitchen after school, then clean up after dinner and take out the trash while his kids watched TV and did homework. He accused me of endless imagined infractions of house rules, then made me strip down to my tighty-whitey underwear so he could swing his belt at full strength across my butt cheeks.

At another place I stayed for the summer, the man of the house made me do hard manual labor each day in the hot sun, then awakened me at two in the morning to make me read aloud from Bible verses describing the perils of being a fool, as in:

The fear of the Lord is the beginning of knowledge, but fools
   despise wisdom and instruction.

(Proverbs 1:7)

The only "wisdom" I learned from those middle-of-the-night Bible lessons was that I was wicked, worthless, and alone.

The streets were scary too, whether in South Central Los Angeles or the South Park and Third Ward neighborhoods of Houston. It seemed like everyone but me had family or someone who had their back. Mama and Daddy were nowhere to be seen or heard, and I refused to join a gang. Joining up with a bunch of violent criminals seemed like a losing proposition since those gangbangers were exactly the shady muthafuckas I was trying to avoid. I felt safer going solo.

So I learned to survive on my own, living by my wits and my fists. I learned that it was best to hit first, hit hard, and to never stop hitting. I learned that if I intimidated the other guy first, he wouldn't intimidate me. I learned to be hard and to act bad to stay safe. Eventually, I *wanted* to be bad. I began to think of myself as the evil villain in the comic book, the guy who drew all his power from being bad.

Through all of the hardships of that time away, teachers were my one lifeline. They fed my intellect and curiosity and made me want to excel academically. They even pulled me into extracurricular clubs, like swimming and math club. But they couldn't keep track of me as I was shuffled from one school to another. And they couldn't protect me from feeling worthless and alone.

Things got particularly dire for me in Houston, where I was living in South Park and taking the city bus to and from school in the Third Ward, where there was a gifted-student program. I had no gang connections or protection, so traveling alone between two warring hoods was treacherous. Just standing at bus stops with a book bag made me a target. After I was jumped twice in the same week, someone called Mama and she got me on a plane back to New Orleans.

WHEN DADDY PICKED ME UP at the airport, I was hoping he'd invite me to come stay with him. Being back with Daddy in his gray pickup made me feel safe, and I didn't want that feeling to end. I was beaming and happy riding shotgun up front and eating sandwiches he'd picked up for us. It was a gut punch when I realized he was driving me straight to Aunt Middy's house on Kelly Hill. When we got there, he told Aunt Middy to find Bridgette in Wesley Chapel and send me to live with her and Dwane. Then he turned around his truck and drove back to New Orleans.

The next morning Bridgette was standing outside Aunt Middy's house. I wanted to run to her, throw myself into her arms, and bawl my eyes out. But the crybaby part of me had gone hard inside. So we just hugged, and I jumped into her car and we drove away.

Bridgette lived with Dwane in a trailer in the woods in Wesley Chapel, halfway between Laurel and Meridian. Mama had moved in with them to tend Bridgette when she was pregnant. But sadly she lost her baby at five months. So Mama had moved back to New Orleans, and I moved into the corner of the trailer where Mama had stayed.

Halfway through breakfast, I realized I wasn't the only person who had changed over the past year and a half. Bridgette talked in a fake Mississippi country accent, and she made me country foods like biscuits from flour and lard, and "syrup" in a saucepan from sugar and water. Once we started eating, I noticed that I was the only one using a fork. Dwane and Bridgette were eating with their hands! I couldn't believe this was the same Bridgette who had always corrected my table habits and who was forever telling me, "Stop smacking!" Now she was doing the smacking—and licking her fingers!

She was acting . . . well, dumb. And I mean acting. Bridgette was nowhere near dumb. In Mississippi they called it "playing country dumb"—usually to trick an outsider. Bridgette and I had both changed. I'd gone bad and she'd gone country dumb.

Even though my new home was just a trailer in the woods, I felt safe there and was happy to be back with Bridgette. Even a country dumb version of Bridgette. Maybe, I thought, I wouldn't need to play it hard and tough back here in Mississippi, living with Bridgette and Dwane. Maybe I could be good.

26

IF YOU WANTED TO GO from bad to good in Wesley Chapel, the road to redemption ran straight through the African Methodist Episcopal (AME) Church. The Wesley Chapel AME congregation was mostly five extended families, including Dwane's family, the Morgans. Right off, Dwane laid down the law: If I was staying in his trailer, he expected me to go to church every Sunday.

Sunday church was routine for me. Mama almost never went with us, but she made Bridgette and me go to Sunday school and church ever since I could remember. Mama was raised Catholic in New Orleans, but she didn't discriminate when it came to church attendance. Wherever we lived, we went to whatever the local Black church was. In Watts we attended Holiness Church, which Bridgette and I called Crazy Church because people were always rolling around on the floor and speaking in tongues and carrying on like crazy people. In Houston and New Orleans we went to Catholic church. I liked the incense, the bells, making the sign of the cross, and volunteering to usher. In Piney Woods we went to East Baptist Church, where the singing was always spirited. When we lived in Pomona we even went to evening Bible study twice a week with Jehovah's Witnesses.

Sin and forgiveness were always the Big Two church teachings, no matter where we went. One of my Sunday school teachers said you couldn't sin until you were seven, but another said you had until you were twelve to know good from evil. The age thing mattered to me, because by the time I turned eight I'd begun dipping my toes into the waters of sin.

On Sundays, Mama would give me some coins from her purse to put in the church collection. But like all kids, I had a deep yearning for candy. Sometimes, on the way to church, I'd peel away toward a corner store to spend my collection-plate money on sweets. Even worse, on days when I gave my dimes and nickels to Jesus, I'd go by the store on my way home and steal a twenty-five-cent Hubig's Pie or fifteen-cent pack of Stage Planks, which was definitely a sin. The next year, when we were living with the Browns in New Orleans, Darin's older brothers taught me how to beat my meat till I got "that feeling" down below. Nothing came out, though, so I didn't know if it counted as sin.

I loved the magic of the Bible stories, and I believed in angels and in God's miracles. Bridgette told me the folks in Crazy Church were shouting and carrying on so much because they'd been "touched" by God. But I couldn't get that "touched" feeling. Why wouldn't God touch me? I wondered. Was it because I had sinned?

Forgiveness was just as complicated. I'd forgiven Now and Later for taking my bike. But if Jesus had grown up in Watts and had turned the other cheek every time some gangbanger messed with him, he'd never have lived long enough to lift up the lepers or multiply the loaves. I couldn't understand that about Jesus. He had the superpowers to work mighty strong miracles, like when he brought Lazarus back from the dead. But Jesus never used his powers to smite his enemies or even protect himself from the Romans.

I had to admit that there were no gangbangers out to get me in Wesley Chapel. I didn't have to prove to anybody that I was hard. If I really wanted to be touched by God, I figured I might need to quit being the tough guy and give meek and mild a chance.

But before I could be touched or redeemed, I was called to work in the woods.

As summer approached, Dwane made it clear I wasn't going to be sitting around doing nothing all day when everyone else over seven years old had to work. Bridgette stocked shelves at Gibson's department store every day, and Dwane worked at a factory making transformers for light poles. My summer job, he told me, would be hauling pulpwood with his father and brothers.

Dwane's father, Ed Howard Morgan, provided for his family by cutting and hauling pulpwood. Six days a week in the summer, his three boys—ages eight to thirteen—would head out into the woods with him at dawn. Their mama, Miss Banny, already had a serious breakfast of grits, eggs, and pork chops cooked up when I arrived at five A.M. for my first morning's work. After breakfast, we piled into the pickup truck and drove down a series of muddy dirt roads where loggers had already cut out the prime trees. Pulpwood haulers lived off the scrap trees the loggers had left behind. Our job was to cut and haul two truckloads down to the lumber mill before the woods got too hot for work.

By the time we returned back to Miss Banny's table for supper that first day, I had broken out from head to toe with tiny bumps on my torso that ran pus and itched like the devil. That night, Bridgette covered me in pink calamine lotion, but nothing seemed to help. Each evening after hauling pulpwood, lying in a tub of water was about the only place I felt comfortable. Miss Banny called it "Indian Fire," but nobody really knew what afflicted me. I felt as though I was at war with my own body. After a week of me battling my skin and losing, Miss Banny told me to take a break from hauling pulpwood.

While everyone else was working in the woods, I was left to myself in the Morgan house. When I searched their house for something to read, the only two books I could find were a giant white Bible with pictures and a standard King James Version. Of course, I knew the Bible stories and much of the scripture from all those years of church and Sunday school. But I'd never read the Bible front to back like I'd done with *Roots* and the World Books. Maybe it was time to finally read the Good Book.

What I'd always liked about the Bible were the miracle and mystery bits—raising the dead, walking on water, and such—which were mostly in the New Testament. But when I started reading the Bible from the beginning, with Genesis, I realized the Old Testament was a family story—a messed-up family full of jealousy and anger and violence. But also love. Not the Jesus-loves-you kind of love, but the romantic kind between men and women, and the complicated kind of love between parents and their kids.

In some ways, the family in Genesis reminded me of my own family, and the families I'd grown up around. Abraham standing over Isaac with a knife. Jacob and Esau hustling and scamming each other for their father's blessing. Joseph's brothers throwing him into a pit and selling him to the Ishmaelites because their father loved him the most.

I wondered if maybe I was the family dreamer, like Joseph—though I didn't feel like my daddy's favorite anymore. I saw a lot of myself in Jacob, who was a dreamer like his son Joseph. But Jacob was also bad, like when he stole his father's blessing and had to flee for his life. Then years later, Jacob returned home and was welcomed back into the family. Maybe Wesley Chapel was my chance to move past my bad, and reclaim my good self. Jesus said, "The kingdom of God is within you." Maybe I just had to look deeper within for the good.

The part of Jacob's story that I read and reread several times was when he wrestled all night with the angel. Everyone knows that angels are good. So why was Jacob wrestling with him? I figured it must be because he wanted the angel's blessing so much. I loved the part where the angel finally gave up, right before dawn, and said, "All right, I'll give you my blessing. You can come home and live here in peace." Ever since Mama sent me out west, I'd been fighting to survive on my own. I wore my tough guy armor for protection, but inside I was still the kid who wanted to play with his sister and curl up with a book. I was tired of all the wrestling. I just wanted to surrender.

Each afternoon when Ed Howard and his boys returned from the woods all sweaty and tired, I was ready to share the Bible stories I'd been reading. While Miss Banny fed us supper, I'd entertain everyone by reciting and even acting out sections of scripture I'd liked enough to memorize. I'd always watch out for scary stories to put a fright into the young'uns, and stories of wonder for Miss Banny. Chris Morgan, who was my age and my best friend in Wesley Chapel, liked the ultra-violent tales where guys got thrown into lions' dens or fiery furnaces. I even tried to get a smile out of sour old Ed Howard by reciting funny proverbs like "Better to live on a corner of the roof than to share a house with a quarrelsome wife." That one got me a smack across the head from Miss Banny.

It wasn't long before my reputation as a Bible teacher had begun to spread across our small community.

27

ONCE THE DEACONS at Wesley Chapel AME heard about the kid in the congregation who was quoting scripture by heart, they got me teaching the kids at Sunday school. As my biblical knowledge grew, so did my teaching responsibilities. By the end of the summer I was leading adult Bible classes too, even though I was only thirteen.

Teaching Sunday school was the first time I stood up in front of folks and recited scripture. Typically, Sunday school was taught from pamphlets an AME publisher sent with prepackaged "morals" and lessons based on a weekly Bible story. But I liked to go off-script from the pamphlets, giving my own slant to the stories, acting out the various parts to bring the drama alive.

Pretty soon, I noticed Reverend Cherry standing in the back of the room, swaying gently with eyes closed and smiling softly as I did my reciting. Reverend Cherry was something unusual in Mississippi: a woman preacher. She was in her fifties and wore a mid-sized Afro with a patch of gray on one side, even though by 1980 most Black women had gone back to perming their hair with chemicals or pressing it with a straightening comb heated on the eye of a gas stove. The women in particular didn't take to Reverend Cherry right off. But Wesley Chapel was a matriarchal community, and over the course of the summer the congregation came to accept and respect her as their spiritual leader.

Sunday church was all about respect, and the old folks had seniority and deference in that department. While the older men
were the deacons who sat up on the pulpit, it was the older women
who ruled the pews. They sat in the front rows, the seats of honor,
dressed all in white with fabulous hats. Many of these women were
prone to shouting aloud each Sunday whenever they felt touched by
the spirit. Often, during particularly impassioned sermons on hot
days, they would require the attention of at least four people to
keep them from fainting, falling, and hurting themselves. So intoxi-
cated were they by the Holy Spirit, they needed a person supporting
them under each arm, while two others fanned them with one of
those ever-present hand fans that doubled as advertising space. "I'm
a fan of Delta Salvage Company," read one. "Your dollar buys more
at the Yellow Front Store," read another. The flip side of the fan
would feature a full-color painting of Jesus on the cross or holding
forth at the Last Supper.

During the Sunday service, I found myself studying the faces
of the older congregants and deacons up on the altar. The rest of
the week, they awoke before sunrise and performed back-breaking
manual labor in the fields or in a factory or in the garden after work.
And all week they swallowed their dignity in the presence of white
folks. The older generation in particular took no chances. No matter how young the white person, they always addressed them with
"Yessuh" and "Yes'm." But on Sundays, it was the older Black folks
who received the Yessuhs and Yes'ms from the rest of the congregation.

Toward the end of one fateful Sunday service—after the doxol-
ogy and the hymns and such—Reverend Cherry strode to the pulpit
to begin her sermon. She looked out across the congregation as if
searching for someone. Then her eyes came to rest on me, and she
smiled. I couldn't tell whether that was good or bad—but I was
about to find out.

"People of the communi-TAY," she began, "I want to share with
you a vision I had last night. As it grew dark outside, I felt a spirit

moving inside of me. I didn't know what it was, but it bothered me all through my dinner. And as I prepared to go to bed, I said to the Lord: 'Lord, use me and tell me what it is I need to know.' So I went to sleep last night, leaving myself open for the Lord's message. . . ." Reverend Cherry seemed to go into a trance right then, closing her eyes and rocking back and forth. "I found myself . . . walking into a building. It was a stately building . . . a mansion."

At the mention of a mansion someone shouted, "Oh yeah, praise the Lord!" Every Sunday we'd hear that the Lord was gonna prepare us all a mansion in heaven.

"I walked into that mansion, and there was a great room, and the room was filled with a beautiful light that called to me. At the end of the room I saw a tall staircase, and something told me to go up that staircase . . . and when I got to the top of the stairs, I saw a figure. The figure had a halo around his head . . . a bright, shining halo of light. . . ."

"Praise Jesus!" someone shouted.

"But the figure's face wasn't clear . . . so I went closer . . . and I noticed that the figure was a man-child. Not a baby, and not a child. Not yet a man, but no longer a little boy. As I drew closer, I saw that the figure was . . . James Plummer Jr."

"Whaaat the—?" I don't know if I said it out loud, but I was sure thinking it loud and clear.

Everyone in the pews turned to look at me, like I was a ghost or a holy spirit. I practically shit my pants.

"And I could see by the shining light of his halo," Reverend Cherry continued, "that this man-child had been called to serve the Lord. I woke up then, but I couldn't get this glorious vision out of my mind. So I asked the Lord, 'Lord, what are You telling me?' And the Lord spoke to me in my heart, and He said, 'This boy has been called to the altar of the Lord.' So right now, I'd like to call James Plummer Jr. up to the pulpit."

I was nailed to my seat next to my buddy, Chris Morgan. He elbowed me hard in the ribs. "You heard the reverend. Get up there, boy!"

Going up to the pulpit was a big deal, because ordinarily kids weren't allowed there. When I finally stumbled out of the pew and up the stairs to where Reverend Cherry was standing, she placed her hands on my shoulders like she was blessing me.

"The Lord has called this young man to the altar," she said, "so we're going to do things a little different this morning. James Plummer Jr. here is gonna deliver today's sermon instead of me." Reverend Cherry smiled down at me, all beatific. Then she stepped back and sat down in the preacher chair behind the pulpit.

Deliver a sermon? I thought. What the fuck?! At first, I couldn't move or talk. I just stood there like a statue in my church-day outfit of pressed bell-bottom slacks, dress shoes, and baby-blue polyester shirt. I could feel the sweat soak through my shirt and a rash start to break out across my torso. I had to call on a higher power not to start in scratchin'.

I had never delivered a sermon or had any intention of delivering a sermon. But I sure had *heard* lots of sermons.

"God is good!" I began, stalling for time.

"All the time!" someone shouted back.

"Let the Lord use you!" shouted someone else.

"Help him, Lord!"

I looked out at the old ladies dressed in white, their hand fans going hard at it. They looked back at me with faces filled with hope, as if God had raised up a prophet in their midst—a man-child called by the Lord to lead them out of the hot, sweaty box of their lives and into the Promised Land. But what could I, Li'l Jame, possibly tell them?

Then I remembered that Samuel was only eleven years old when he was called to be a prophet. God called to him three times before Samuel responded. Was the Lord calling to me, I wondered, through Reverend Cherry?

I turned to look back at the old-ass deacons sitting behind me. Every one of them looked two hundred years old, easy. A couple of them could have been in the Bible. Methuselah Jr. Noah the Fifth. Then I turned back to face the ladies in the front pews. Looking at all their old, crinkled faces reminded me of the prophet Ezekiel in the Valley of the Bones—a Bible passage that spooked the hell out of me ever since I first read it on the Morgans' living-room floor. That's when it came to me: these old men and women wanted to be delivered out of the Valley of the Bones and into the Promised Land.

"The hand of the Lord was on me," I recited from Ezekiel, pausing for effect as I cast my gaze on the old ladies in white. "He set me down in the middle of a valley, and it was full of bones."

"Tell it, young one!" a lady in white shouted back.

"I saw a great many bones on the floor of the valley, bones that were very dry. And the Lord asked me, 'Son of man, can these bones live?'"

"Yes, they can!" the congregants called in response.

"This is what the Lord told me. He said, 'I will put breath in you, and you will come to life! Then you will know that I am the Lord.'"

"*Praise* the Lord!" said an old lady, rising to her feet, fanning like to put out the fires of damnation.

"Then He said to me, 'Come, breath, from the four winds and breathe into these slain, that they may live.' And breath entered them, and they came to life and stood up on their feet—a vast army."

"Weeellllllll!" someone shouted.

"That's a'right!" someone else called out.

Folks in the congregation began to get to their feet, like a vast army of the dead arising. I was working up a head of steam now, putting some singsong into it.

"Then, he said unto me: 'Son of man, these bones are the people of Israel. Therefore, prophesy and say to them: My people, I am going to open your graves and bring you up from them; I will bring you back to the land of Israel. I will put my Spirit in you, and you will live, and I will settle you in your own land.'"

"Hallelujah!"

"Praise the Lord!"

I knew I really had them going because even the old-ass deacons were creaking to their feet. And all I had done was singsong some Bible verses I'd memorized one morning so I could scare the daylights out of little Eddie Morgan when he got back from hauling pulpwood. But you'd have thought I had written this scripture myself by the way the old ladies were rocking back and forth and fanning themselves to beat all.

The fear faded from my body as I began to feel the power of my preaching. I could see that folks suffered. They suffered with illness. They suffered with poverty. They suffered with concern for their children and loved ones. And they needed hope. I could give them hope that another, better world awaited their tired old bones. I didn't need to be funny or intimidating. I could inspire and lead these people. I could be good, and do good.

"Praise Jesus!" someone shouted, urging me to keep the hope flowing.

But I also had a secret that held me back: I no longer believed as they believed. I no longer believed any of it.

During my summer studying scripture, I'd run smack up against some serious cosmology problems, starting with God creating the world in seven days. When I asked Bridgette and Dwane what was up with that, they became suspicious. "Why are you asking that? Are you an atheist now?" Ever since I'd encountered the gospel according to Einstein, I considered myself a man of science. I needed evidence before I could believe something as important as how the universe was created. Praising and believing in the Lord wasn't enough for me. I needed to *know*. And to know, I needed tangible, reproducible evidence.

By the time I'd finished reading all the miracle-making I'd loved as a kid—from creation all the way through Jesus rising from his grave on the third day—the Bible just seemed like a storybook some folks had made up, no different from a superhero comic book or a Greek myth. I wanted to feel touched. I wanted to be loved and redeemed. But I didn't believe anymore in the Kingdom of God, inside or outside of me.

Meanwhile, I was still standing at the pulpit, and the old ladies in white were still standing and swaying in the humid heat. I'd wound them all up, and now I had to wind 'em down. Having grown up in the church, I knew all the tricks of the trade. But I was suddenly out of tricks.

I was sweating furiously, so I grabbed a hand fan from off the podium and began to fan myself. That's when I saw it, as if God's plan had been delivered into my hand. One side of the fan had a painting of Jesus in a red flowing robe. The other side was a funeral-home advertisement that read: "When the shadow falls, who you gonna call? Benson's Funeral Home: 601-334-4400."

"When the shadow falls," I said aloud, "who you gonna call? . . . Jesus, that's who!"

"Preach it, young one!"

"When the breaks go against you, are you gonna break? Or are you gonna call Jesus? 'Cause these here breaks will rock your shoes."

"They sho' will!"

"If your woman steps out with another man, that's the breaks. If she runs off with him to Japan, that's the breaks."

I was getting my swaying and singsong groove on now, and everyone started to move along with me. Except for Chris Morgan's mama, Miss Banny, who was standing right up front and pressing her fists against her mouth like she was gonna explode. Just last weekend she'd heard Chris and me singing and dancing to the new Kurtis Blow song, "The Breaks," which I was now lifting from like a bandit.

The whole congregation was on their feet. I stepped out from behind the podium and started mixing in some breakdance moves I'd been working. First, I'd pop, then lock, wave, pull the rope, and do the shake. I started with one hand shaking, then stuck it into my pants pocket and started my leg shaking. Then I sent the shake up my body until all of me was shuddering and quivering as if a thousand amps of electricity was flowing through me.

"Throw your hands up in the sky, and wave 'em 'round from side to side," I shouted. A hundred hands shot upward.

"And if you deserve a break tonight, somebody say all right!"

"All right!"

"Say ho-oo!"

"Ho-oo!"

"And you don't stop. Keep on, somebody scream!"

"Owwwww!"

"Break down!"

As if on command, one of the fan ladies in the front row—a big, wide one with a floppy white hat—started to fall into a swoon. No one was there to prop her up, so I leapt off the pulpit and landed right in front of her. She tilted to one side, then the other. When she tilted forward toward me, I caught her with both my arms outspread, just like I'd seen real preachers do.

I called out to the Lord, like Samson before the Philistines, "Please, God, strengthen me just once more!" But the Lord must not have heard my plea. Or else He knew that *I* was the Philistine, not a servant of the Lord like Samson. My legs buckled under the enormous weight of the fan lady, and I collapsed beneath her.

Not the big mic-drop finish I was going for. But my first, and last, sermon was over.

28

AFTER MY FALL FROM GRACE at the AME, Good James and Bad James resumed their wrestling match.

In seventh grade at Southside Middle School, the nerd in me still wanted to be the smartest kid in the class. But they set the bar so low in backwoods Mississippi that you could step over it with your eyes closed. I mean, school at Southside was barely school.

On my first day, when the bell rang at the start of social studies, a group of girls jumped up and started singing Peaches and Herb's "Reunited," moving in elaborate synchronized dance moves they all seemed to know. Not to be outdone, a cluster of boys started pitching pennies at the back of the room, while another group of guys formed a circle on the floor and began shooting marbles. The teacher, meanwhile, put his feet up on his desk, unfolded a newspaper, and read the sports page.

I sat at my desk, dumbfounded. After a few minutes of watching the girls sing and dance, and the boys pitch and shoot, I decided to open the textbook and read about religious intolerance in colonial America.

By third period, I'd already made a reputation for myself as the "book-smart" kid at Southside because I could read out loud just as easily as I talked. The English teacher went around the class asking kids to read aloud from a book. If I hadn't known better, I'd have thought it was first or second grade. That's how bad my classmates were at reading. I couldn't take any pride in being the best reader in the class when no one had even bothered to teach the other kids to read.

Outside of class, I carried myself with an air of volatility. I talked mad shit and tried to be as outlandish as possible, just to keep folks off guard. I figured that if I acted menacing and unpredictable, older guys would back away rather than attack. There were no gangs in Wesley Chapel, but I figured I was better off dressing urban and acting tough anyway. My favored outfit was white leather high-tops, a pair of black baggies, and a white button-down shirt with suspenders.

I had only two fights that year. But if you came out on the right side of them, two fights were enough to send a message. After Willie Earl sucker-punched me on the bus, I beat him down good in the back row of seats. That established my reputation as a guy not to mess with. I got suspended for a week and spent it exploring the woods behind Dwane's trailer, following Reedy Creek to see where it would take me. That's where I stumbled on someone's weed grow.

There were more than twenty plants, each one around four feet tall, spread out over a wide area. I decided to pick leaves off of several different plants so the owner wouldn't notice they were gone. I put the weed into a brown paper bag, folded it up, and stuffed it into my back pocket. Within two days, it was dried and cured.

I found a pack of Tops rolling papers in the back of the junk drawer in the trailer—but I didn't know how to roll. I asked Chris Morgan, but he didn't know how either. So we decided to go to the Hole, which was a spot a mile up the road where the grown men would hang out and get fucked up on malt liquor and fortified wine. An old guy named PeeWee rolled my weed into four joints and kept one as payment for his labors. Watching him roll those four joints was all the schooling I needed.

I became a daily weed smoker after that, and Chris became my regular after-school smoking partner. We never had homework, so after classes we'd duck into the woods and fire up. Chris and I also started sneaking off to juke joints and pool halls on weekend nights. We were the only seventh graders in those clubs. But just like at the B&M, no one ever checked our IDs. And since I had a virtually endless supply of weed from the grow I'd found in the woods, I could always trade someone a joint for a fifth of T.J. Swann.

Smoking weed marked the start of my double life, which required a steady supply of Visine and breath mints. If Bridgette or Dwane found out I was smoking weed, I'd be in deep shit. So I was super careful to never smoke near the trailer or even keep a stash there. At home, I was still Li'l Jame and the hierarchy between Bridgette and me was the same as always. I wouldn't dream of giving her sass or talking back to her. Ever. Which made things simple. It meant that when I was home in the trailer, I could park the cool, urban smart-ass and curl up with a good book.

BACK AT SCHOOL, I'd become a disrupter in the classroom. I couldn't resist being the class clown and cutup. Part of me still wanted to be the guy with all the right answers when the teacher asked questions, but it was hard to take them seriously enough to behave respectfully in class. Half the time the teachers were teaching stuff wrong. And when I corrected them, they immediately bounced me out of class and sent me to the vice principal's office.

I was also butting heads with the sports coach. All the able-bodied boys at Southside were expected to play football. Particularly someone like me, who at thirteen was already as tall and broad as a grown man.

"The teachers say you a smart boy," said the coach the first day of practice that fall, "so we're making you the quarterback."

I didn't want to be a quarterback—at least, not with hard tackling. I had loved playing touch football in the streets of New Orleans with Darin, which was a finesse game. The whole idea of tackle football seemed stupid and dangerous. Especially since there weren't enough helmets to go around, and we had to take turns wearing them.

Before the season even started, I was regularly cutting practice, running the halls, or smoking cigarettes behind the school. One day I was chilling behind the gym and getting ready to light up a smoke when the vice principal, Mr. Jones, came around the corner.

"James Plummer," he called out before he even saw me, as if he already knew I'd be there. "Come with me, boy."

Mr. Jones was well over six feet tall, and he carried himself with authority. I assumed I was in for a paddling, since Mr. Jones was known at Southside as the paddler-in-chief. But instead of heading to the main office, Mr. Jones walked me up the hill at the back of the school that led to the band hall, a building I'd never been inside.

A couple dozen kids were practicing on woodwind and brass instruments in the main open room. In an adjacent office, sitting behind an old wooden desk, was the band director. Mr. Cross was a dark-skinned man of short stature with a thick mustache and a tobacco pipe securely clenched in his mouth. He was busy writing musical notations on a piece of paper covered with musical staves from top to bottom.

When Mr. Jones knocked on the open door, Mr. Cross looked up slowly from his page.

"Mr. Cross. Good afternoon, sir."

"Mr. Jones," Mr. Cross said. "How may I help you?" Mr. Cross leaned back in his chair, struck a match over the bowl of his pipe, and inhaled a few times, sucking in and exhaling a cloud of sweet-smelling smoke.

"Well, we have a bit of a problem that you might just be the man to solve," Mr. Jones said. "This boy here is James Plummer. His teachers say he's smart but he's starting to go down a bad road. For the last few weeks he's been running the halls every day instead of going to football practice. We need to find something for him to do to keep him straight."

Mr. Cross got up, walked around his desk, and looked me up and down. He relit his pipe and sucked on it, as if considering whether I was worth the effort to straighten out. "Leave him with me, Mr. Jones. I'll put him to good use. And I'll teach him some discipline while I'm at it."

"Thank you," Mr. Jones responded. "If he can't be straightened out by you, we'll have to take more serious actions."

Mr. Jones left, and I was alone with Mr. Cross.

29

M<small>R.</small> C<small>ROSS LEANED</small> on the edge of his desk and squinted at me through the smoke curling out of his pipe. "What is discipline?" he asked me.

"It's when you do something you don't want to do," I replied. "But you do it anyway because you supposed to. That means you got discipline."

Mr. Cross shook his head. "Discipline is the training that makes punishment unnecessary. I'm going to teach you discipline. Either you'll learn discipline, or you'll receive punishment."

He reached into his desk drawer and handed me what looked like a little silver funnel. He took out another one and put it to his lips. "Do this," he said, and made a buzzing sound. I attempted to mimic him but only blew air silently through my mouthpiece. Calmly, he removed the object from his mouth, pressed his lips together tightly, and blew air through them, making a buzzing sound. I did the same, buzzing my lips. Then he returned the mouthpiece to his lips as he continued to buzz, making the same sound he made earlier. I mimicked him, bringing the object to my lips and buzzing as he did. To my amazement, I could now make the sound, but just barely.

"This is a trumpet mouthpiece," he said. "Let's try these others."

He pulled out two more mouthpieces. One was twice the size of the trumpet mouthpiece and the third one was about twice the size of that one. He asked me to buzz each of them in turn. I could barely get a buzz out of the two smallest funnels, but I was able to easily get a buzz out of the largest one.

"That's for a tuba," he said. "Sounds like you've found your instrument."

Mr. Cross directed me to change the pitch of the buzzing sound by pressing my lips together more tightly for a higher pitch or more loosely to create a lower-pitched sound. He then directed me to touch the tip of my tongue to the ridge behind my upper teeth so that I could break up the sound with each touch. He then counted and told me to tap my foot in time to his.

"One, two, three, four; one, two, ready, play . . ."

Buzz, buzz, buzz, buzz; buzz, buzz, buzz, buzz; on every footfall.

"From now on, you come to the band hall at fifth period every day. What you will do is practice buzzing on that mouthpiece and keeping time. I don't want to see you out here talking or grab-assing. You sit in this chair and you buzz on that mouthpiece keeping time with your foot. Do you understand?"

"Uh-huh." ·

"You want to try that again? Do you understand me?"

"Yes," I said, looking him in the eye.

"Yes what?"

"Yes, sir, Mr. Cross."

"That's better. From now on, you address every adult as sir or ma'am. Do you understand me?"

"Yes, sir."

"What is discipline?"

I didn't remember what he'd said, so I just looked at him.

"Discipline is the training that makes punishment unnecessary," he repeated.

FOR THE NEXT WEEK all I did was buzz on that damned mouthpiece while all the other band students looked at me and giggled. After a week of me buzz-buzz-buzzing, Mr. Cross called me into his office. Next to his desk was a sousaphone—which is the marching-band version of a tuba—sitting in an open case. It looked brand-new.

"Go ahead," he told me. "Pick it up and try it on."

It was heavy. Trying not to grunt with the effort, I lifted it overhead and lowered it down around my torso.

"It looks good on you," Mr. Cross said, stepping back and nodding his approval. "I've been waiting a while for a tubist to walk into this building. Put the mouthpiece on."

I attached the mouthpiece and I buzzed into it just as I had done so many times before. But instead of the tinny, high-pitched sound, a loud, booming bass issued forth from the tuba's bell. I was startled by the authority of its voice. When the note vibrated through the tangle of brass tubing, I felt it resonate throughout my whole body. It was loud and powerful. And it sounded like music!

After teaching me a few notes, Mr. Cross seemed pleased. "Here is an instruction book. I want you to play from it each day. You'll learn new notes, how to count, and the main keys. It also has scales and arpeggios. Memorize them and practice them every day."

I BONDED WITH MY TUBA that year. Trumpets and trombones were pretty cool, because they play the lead voices. But bass drives marching bands—particularly Black marching bands. The bass from the tuba and the drum never stop. They move the crowd. You may hear the lead, but you *feel* the bass. I had a new purpose and a new persona. Tubas were bad. I was bad. We fit.

Music became my new passion. I quickly learned how to read music—not just for the tuba but for all the instruments. Just reading the sheet music, I could feel the combined power of brass, woodwind, and percussion singing out in one amplified voice.

I progressed so quickly that Mr. Cross invited me to play the next year with the Heidelberg High School marching band, where he was also band director, even though I'd still be a middle-school eighth grader. I leapt at the offer, even if it meant practicing during the heat of the summer. The first time I had to wear a thirty-pound sousaphone all afternoon in the August heat I felt so dizzy I thought I might puke or pass out from heatstroke. But I didn't care.

One afternoon in August as I was leaving band practice, a horn beeped behind me. It was Mama sitting at the wheel of a car I'd never seen before. A brand-new beige Datsun 210. A Japanese car in Clarke County in 1981 was about as uncommon as Halley's Comet. You never could tell what was coming next with Mama. By then, I'd long ago learned to roll with her surprise moves.

Six months earlier, she'd appeared from I-don't-know-where and moved into Dwane's trailer with us. Dwane helped Mama get a job on the assembly line where he worked at Howard Industries, and he loaned her a beater of an old Vega to drive. I guess she'd saved up some money in the past few months, because that Datsun looked brand-new.

I was playing with the radio dial when Mama took a left at Main Street through downtown Heidelberg instead of keeping straight on Pine Street toward Wesley Chapel. I figured maybe she was going to the liquor store next to the railroad tracks. But she crossed the tracks, took a left toward Piney Woods, and headed up a dirt road not far from Aunt Middy's farm. She brought the Datsun to a stop in front of a new brick house.

"We visitin' someone?" I asked.

"This is our new house, baby. Your stuff's in back," she said, motioning toward the trunk. When I gave her a puzzled look, she squealed out loud, "My FHA loan came through, baby. We homeowners now!"

Before I stepped foot inside our new house, I wanted to know if this move meant I'd be switching schools again, just when I'd made band. Mama said not to worry, that she'd drive me to the West Brothers store on Highway 11 where I could catch the Heidelberg High bus coming out of Vossburg.

Inside the house there was new furniture she'd bought on credit. And something else: a brand-new set of Encyclopedia Britannica—not the World Book, but the real deal—all lined up like a proud troop of burgundy soldiers inside a new wooden bookshelf. My own private army of smart.

30

EVERY KID LOOKS FORWARD to moving up from middle to high school. But when I arrived at Heidelberg High, I felt like I'd taken the elevator from the basement to the ground floor.

Back in the 1980s, Mississippi seemed to compete with Alabama to see which state could spend the least amount of money on education, and whose students could score the lowest on standardized tests. In 1982, Southside Middle School and Heidelberg High were ranked as the second- and third-worst-performing schools in Mississippi.

Truth be told, Heidelberg High wasn't much more than a shell of a school. When the courts ordered Mississippi to integrate its public schools in the mid-'70s, the local whites burned down "their" high school rather than share it with Black kids. Then they started Heidelberg Academy as a private school for whites, and the state had to build a new high school to replace the one they burned down. So the Black kids got a new building, but Heidelberg High had to raise up its administration, faculty, and student body from the ground level.

Heidelberg High's student body was all Black, and so was the faculty. All except for my homeroom and civics teacher, Mr. Reeves. He was short and fat and the frequent target of jokes and pranks. But he was the only teacher who consistently had my back. Not surprisingly, some of my teachers resented my disruptive classroom behavior. Since I read the textbooks the first week of the term, I tended to sit in the back row of class, cutting up and teasing girls. I often understood the course materials better than the teachers did, and I didn't take the trouble to hide it. So I was sometimes considered an obnoxious troublemaker who needed to be controlled, or just kicked out.

Mr. Reeves appreciated the intellectual energy I brought to the classroom. He believed in me, and more than once he came to my rescue—like the time I got caught launching bottle rockets off the roof of the band hall during school assembly. "He's a mischievous kid, but he's a brilliant mischievous kid who we need to encourage," Mr. Reeves told the principal. "There's nothing really wrong with James. He's just bored."

He had that right. I was bored half to death. It was a daily battle for me just to stay awake in classes. Then one day in the middle of the school year, fate intervened to call a cease-fire in my war with boredom.

THE WEEK AFTER we returned from Christmas vacation, two professors from the University of Southern Mississippi showed up at school to try to convince us to compete in the regional science fair USM was hosting that spring. These dudes did not look like anyone we knew or would become. They were even more white and more square than Mr. Reeves.

Most of the kids just stared out the cafeteria windows while the two white professors talked about the mind-blowing projects other students had created for science fairs—like a robot that walked like a dog. We kids at Heidelberg High had never even heard of science fairs. Telling us that we could compete against kids who built robot dogs was like telling us we could grow up to be president of the United States. It was technically possible, but too far-fetched to even imagine.

After the assembly, no one mentioned the science fair again. Not even our science teachers. Then, a few weeks later, a magical and mysterious gift arrived from the marketing department of IBM: a spanking new 5150 Personal Computer. At least, that's what it said on the side of the big white box that sat on the floor of the science lab. No one seemed to have any idea what to do with it. After a few days, our science teacher, Mr. Barber, pulled the computer out of the box and put it on the black countertop. Nobody bothered to turn it on, or even plug it in. It just sat there in stony silence like the slab in *2001: A Space Odyssey*.

During lunch period, I found Mr. Barber in the science lab eating a sandwich and reading the newspaper. When I asked him if I could turn on the computer, he said, "Go for it. Just don't break the damn thing."

WHAT MR. BARBER didn't know was that I'd already been experimenting with computer programming for a few weeks. The ads for the first home computers that had begun running on TV that fall were hard to relate to—smiling white families playing computer games and serious-looking white couples calculating their mortgage payments. But then, the day after Christmas, I encountered an actual computer at my friend Anita Page's house.

Anita was a smart girl in my homeroom at Heidelberg High who also played bells in marching band. Anita's mom was a teacher and her dad had a good factory job, so they could afford luxuries that were out of reach for most families in our area—like the brand-new Tandy TRS-80 color computer sitting at the center of their dining-room table.

The TRS-80, Radio Shack's answer to the Commodore 64, wasn't much to look at: a tiny TV screen sitting on top of a CPU and five-and-a-quarter-inch floppy disk drive, with a toy-sized chiclet keyboard. But watching the cursor blink on and off like a faint yellow heartbeat, I could sense some sort of intelligence at work inside its processing unit.

After Anita and I played Pong for a couple hours, I got bored and restless. Inside the instruction packet I found a small stapled book titled *BASIC,* which stood for Beginner's All-Purpose Symbolic Instruction Code. Before New Year's rolled around, I'd taught myself how to program in BASIC.

Programming was like learning to read musical notation. BASIC was a secret language that no one for miles around understood except me and the TRS-80. Using BASIC, I could talk to the computer and give it commands. If I could state the rules in BASIC, I could train the TRS-80 to play simple games. I wondered what else I could teach it to do.

ONCE I BEGAN WORKING on the IBM 5150, I quickly realized it was more powerful than Anita's Radio Shack model. I spent every break period in the science room working on the IBM. Sometimes after band practice I'd sneak into the school and hack away at my programming skills until the janitor kicked me out. I needed to get a computer of my own.

I pleaded with Mama until she bought me a used Texas Instruments TI-99/4 desktop computer at Hezekiah Salvage in Laurel. Hezekiah Salvage was pretty much our one-stop shop for everything in those days, used or "almost new." I set up the TI-99/4 on our kitchen table and wiped it down lovingly with some Clorox, just to clean the nastiness off it. Then, I worked at the keyboard every minute I was home, unless I was practicing on my tuba or sleeping.

Late one night while programming some classic chess openings, I began wondering what was inside the computer. Without turning off the power, I loosened the row of small screws I found on the bottom and back panels and lifted off the outer shell. I stared in wonder at its electronic innards: an inscrutable mash-up of black memory chips, orange capacitors, and striped resistors, all eerily quiet even while working.

I don't want to call it a religious moment, because I was a committed man of science by the time I got to high school. But it made me think of how Moses must have felt when an angel of the Lord called out to him from the burning bush that didn't consume itself. I didn't exactly hear voices speaking to me out of the disassembled TI-99/4, but there was something almost holy about the feeling that came over me. The deeper I stared inside the computer, the more I felt the presence of a bigger, more connected intelligence—an intelligence that didn't seem to dwell inside the machine so much as flow through it.

When I tried to guess what Einstein would have done with a computer, I imagined it would have something to do with relativity. My Britannica encyclopedias had taken me a lot deeper into quantum physics than the World Books, with longer, more detailed articles on special and general relativity, and spacetime. I found citations at the bottom of those articles and asked Mama to bring me whatever she could find at the public library near her factory in Laurel. But book learning could only take me so far. I yearned for a personal experience of relativity, and I hoped the computer might be my portal to spacetime.

As I stared at the TI-99/4's motherboard, I wondered: How do computers do calculations so fast? If I could shrink myself tiny enough to crawl inside one of these microchips, I could probably witness electrons moving near the speed of light. And that's when it came to me: computer games were based on mathematical models—just like Einstein's relativity! If I could translate the equations of special relativity into BASIC—time dilation, length contraction, relativistic mass, and spacetime intervals—I could create a computer program that could simulate relativity and let me experience spacetime!

I must have been in some kind of trance, because I didn't even notice that Mama had returned home from her late shift at the factory and was standing right beside me.

"Boy! You done took the thing apart? Lord, I spent good money on that machine. If you break it, I'ma break yo' ass."

I snapped out of my spacetime stupor and reassembled the computer as fast as I could. Then I typed a quick set of commands on the keyboard and called out, "Ma-muh! Come look at what I wrote for you." She came back in, a Dell puzzle book in hand. "Watch this," I said, and pressed the Enter key.

Line after line of JAMES PLUMMER JR LOVES HIS MAMA streamed down the screen in a continuous scroll. Mama's face split open with a smile. "Shit-ass! You ain't gonna sweet-talk me. Do yo' schoolwork like you supposed to. And clean up them damn dishes outta the sink!"

A WEEK LATER during lunch, I found Mr. Barber in the science lab and told him I wanted to enter the science fair.

"Enter what?" he asked, like he'd forgotten all about it.

"The regional science fair USM is hosting in Hattiesburg. I've been working on a computational project."

"Look who's throwin' around four-syllable words this morning," he said with a chuckle. "You're always showing off."

"Computational" was five syllables, but I didn't want to sass Mr. Barber. Not today. So I just told him, "I'm developing a program to model the effects of special relativity."

That got his attention. He put down his sandwich and newspaper and looked at me side-eyed. "You can do that?"

"I think I can. I've already started. I've learned to program in BASIC and I've broken down special relativity into its mathematical components. It ain't as complicated as it sounds. I been studying the equations since I was eleven years old and I understand 'em pretty good. Programming them is the easy part. What's gonna make my project fresh though is that I'm programming it into a computer game. I can show you on the IBM."

"Hold on, boy," Mr. Barber said jumping to his feet. "Let me go get the other science teachers so they can hear this too." Mr. Barber went to fetch Mr. Dubose and Mr. Atterberry, the biology and chemistry teachers.

The three of them sat on tall stools finishing their sandwiches, while I explained Einstein's thought experiment that led to time dilation. Then I walked them through how I planned to program the equations of special relativity in BASIC and then import them into a game interface.

They had all heard of Einstein and relativity, but only Mr. Barber, who studied physics in college, truly got it.

I could tell I was losing the other two when I explained that time-dilation equations could illustrate how two events that are simultaneous to a stationary observer would be nonsynchronous to an observer moving at high speed. So I typed in the length-contraction formula to illustrate how the two observers would see distances differently.

"So you see, there is no such thing as simultaneous events. Time is fake!"

They weren't seeing it, so I just rolled into my big finish. "Here's the cool part," I said, getting all jumpy the way I did when I got hyperexcited. "If you combine space *and* time like this . . ." I said, typing commands on the keyboard, "you can calculate the invariant interval . . ." I pressed the Return key and the text appeared in perfect equation format:

$$ds^2 = c^2dt^2 - dx^2 - dy^2 - dz^2$$

". . . this equation combines space and time. So even though different folks will observe different times and distances, they all see the same value for the combined interval. That's what Einstein called spacetime!"

Mr. Barber looked at Mr. Dubose, who looked at Mr. Atterberry, who looked back at Mr. Barber. Mr. Barber looked at the computer screen, and then at me. "Okay, so you want to enter this in the regional science fair," he said. "How soon is that?"

But Mr. Atterberry was having none of it. "You ain't as smart as you think you is," he said to me, slowly shaking his bald head surrounded by a nappy fringe of hair and beard. "I took physics in high school, and they didn't say nothin' about no spacetime interval. I think you made that up."

I ignored Mr. Atterberry and focused my attention on Mr. Barber. "The regionals are in six weeks. If I'm gonna get everything done in time, I'll need to take the IBM home with me so I can work on it at night. I can't program the game part on the TI-99/4 my mama got me."

"Why should we let him take this computer home when I hear he's selling weed right here on campus?" Mr. Atterberry said to the other two teachers. "I don't trust this boy."

"I promise you, Mr. Barber, you let me take the IBM home, and I'ma put Heidelberg on the science map in Mississippi."

The science teachers all looked at each other again. Mr. Dubose figured it was his turn to chime in. "It'll be our ass on the line if he messes up the school's only computer."

"After Mr. Cross let me take a brass upright concert tuba home to practice," I said, "I made all-state band."

In the end, Mr. Barber talked to the principal, who called Mr. Cross to make sure I hadn't fucked up the tuba or pawned it, or whatever they imagined I might do with the computer. Mr. Cross reported I was showing a lot of progress in the discipline department.

The next day, Mama drove up to school, and we loaded the IBM into the back of her Datsun.

31

CONVINCED MY TWO NERD PALS, Antrum McGee and Aristotle Bender, to come up with science-fair projects too. Antrum was as country as anybody, having worked with his daddy from an early age on all manner of farm equipment. He designed a solar water heater from sections of a car radiator and other scavenged car parts. Aristotle, who also grew up working a family farm, built some sort of battery powered by manure, which I didn't even want to understand.

We barely made it past the regionals. Not because of our science, but because our displays were bush-league. No one told us that presentation accounted for almost half your score. Everyone else had trifold boards made of actual wood connected by brass hinges. Our boards were made of flimsy pieces of cardboard patched together with tape, with hand-lettered notebook pages pasted on top. Their exhibits each included an "abstract" and a "hypothesis"— words we'd never heard of.

There were plenty of Black schools in the region, but we were the only all-Black team competing at Rebel Coliseum on Confederate Drive in Hattiesburg. So we didn't just look pathetic. We looked like pathetic Black kids playing way out of our league. We assumed the judges took pity on us since we each placed in the top three in our categories—which allowed us to advance to the state science fair in Jackson.

Now that we knew the score, we wouldn't make the same mistakes at the state fair. Mr. Barber took us down to Laurel to get wooden boards cut to the maximum size allowed. We saved up money to buy the boards, hinges, construction paper, stencils, colored markers, and rubber cement—the glue of choice for science-fair projects.

Another thing we learned from the regionals: we needed color—lots and lots of color. If your project didn't look great it was impossible to win. I wasn't much of an artist, but lucky for me, Mr. Reeves stepped up big-time to make sure my graphs and exhibits looked good. By the time I was finished, I had three-color plots for mass, length contraction, and time dilation set against colored construction paper to paste up on the boards. And the title of my exhibit—"Programming the Effects of Special Relativity"—was bright and bold enough to read from fifty feet away. Best of all, the principal allowed me to take the IBM with me to the state science fair in Jackson so I could demonstrate my relativity modeling in real time.

BEFORE DAWN ON the morning of the state science fair, Mr. Barber loaded our boards and exhibits into a van and picked us up from our homes scattered across Jasper and Clarke counties. I came out of my house cradling the IBM in my arms and carried it on my lap all the way to Jackson. Antrum and Aristotle napped the whole way, but I was much too nervous and excited to sleep. I'd been up half the night running and rerunning my simulations on the IBM.

When we arrived at the Jackson State basketball stadium, I woke up Antrum and Aristotle. We went around to the back of the van to unload our boards and the folders with the diagrams we had to paste up on them before the fair started at ten. But there were only two sets of trifold boards in the back—Antrum's and Aristotle's. Somehow, Mr. Barber had left mine behind. I felt something crack inside my chest.

Aristotle handed me his boards. "Here, take mine," he said.

"What? I can't do that," I said. I walked away from the van and sat down on the grassy slope outside the stadium lawn, dejected and disappointed, but resolved to take the loss like I imagined a manly man would. I told myself it was part of being disciplined.

Aristotle sat down next to me. "You gotta take my boards, man. You're our best shot. My battery couldn't power a toothbrush, and everyone knows it ain't gonna win shit. You gotta go in there and represent us."

"Nah, I can't do that. You worked hard on your exhibit, and you deserve to be in there," I said. And I meant it.

"Plummer, I'm not asking you. I'm *telling* you. Take the damn board. Paste your shit up on it and go win this damned thing." He left the board at my feet and walked away.

"Well, hell," I shouted after him. "You gonna help me paste up my boards, or what?" But he was gone.

When I opened my folder, the wind almost whipped my papers away from me. I placed a rock on the folder to keep it closed. While I worked to glue my title banner to the board with rubber cement, the wind blew open my folder and scattered my plots and graphs onto the lawn. I chased them down and returned to pasting them up.

"You really are working hard to get that done," said a woman's voice. Seated just downhill from where I was working, I saw a white husband and wife team who were chaperoning their group of about twenty white students. They were all eating individually wrapped sandwiches and drinking sodas with straws.

"Yes, ma'am," I replied. "This wind is trying to mess me up. But it can't stop me."

Just then a gust of wind scattered my exhibits across the grassy slope. The kids in their group rushed after my papers and retrieved them. They were only a little beaten up. When the woman teacher noticed the title of my project, she said, "That's very impressive!" in a tone of voice I recognized from white teachers I'd had going back to first grade. It was meant as a compliment, but came off more like a put-down, like what she was really saying was, *How impressive that a poor Black boy like you would have such a big white idea. You must be one of those exotic smart Black kids I've heard exist.* I thanked her anyway, hearing Dwane's voice in my head threatening to kick my ass if I was disrespectful to an adult.

When I was finally finished pasting up my boards, I called over Aristotle and Antrum to take a look. They dapped me up and the white folks clapped in polite applause, as if just pasting up my boards was worthy of a prize.

32

ONE LOOK AROUND the cavernous Jackson State basketball stadium told me we weren't at the regionals anymore. The number and quality of the exhibits were on a whole other level. There were aisles upon aisles of exhibits with banners at the end of each aisle telling you what section you were in: Biology, Chemistry, Mathematics. My booth was in the Computer Science aisle, in between a guy who had used his Commodore 64 to calculate corn-planting and -harvesting dates based on daily temperatures and rainfall, which was pretty cool, and another guy who had created a computer game that let you "shoot" animated tin cans off a fence post, which was slicker than anything I'd seen before on any computer. All I had was plots and curves and text—he had real graphics!

At noon, the judges started walking up and down the aisles with clipboards and score sheets. These weren't grad students, like at the regionals. They were actual college professors dressed in short-sleeved shirts and neckties with "Judge" badges hanging around their necks.

Almost immediately, a group of judges clustered around my exhibit, which I figured from the regionals was a good sign. Then, one of them came up to me looking almost embarrassed. "I have to tell you, we're computer scientists, and we have no idea what your special-relativity exhibit is about. We're moving you from Computer Science to Physics."

A half hour later, half a dozen physics professors were examining my exhibit over in the Physics section. One thing they didn't tell us about the state science fair is that part of the scoring is based on how well you can explain your project to the judges. I hadn't prepared for questions, and I was nervous as hell to be surrounded by a bunch of white physics professors quizzing me about special relativity. But after I responded to the first few questions without missing a beat, I realized that I'd been preparing for this conversation ever since I was ten years old. I'd always figured there must be someone else out there who knew enough about relativity to actually discuss it. Now I had six of them at once, and they were all talking about my program. Who cared if they were square white dudes with pen stains on their shirt pockets? They spoke my language!

One of the judges tried to stump me with a trick question: "What happens when you go faster than the speed of light?" I looked back at him with a "That's-your-best-shot?" face and said, "C'mon, sir, that's super-easy math. One minus v-squared over c-squared becomes negative. Can we talk some serious physics?" After that, the questions got tougher. They were probing to find out what I knew and what I didn't know about math, physics, and programming. But my shit was tight. I was loving it.

"Where did you get this idea?" one of them asked.

"From Einstein."

"No," he said, smiling, "I mean, who gave you the idea to program special relativity? And who helped you with the programming?"

"Nobody. I taught myself about special relativity, and how to program in BASIC." I decided to skip the part about my mystical moment staring at the insides of the TI-99/4. "Then I figured out how to program the input and output processes to make it like a game."

After studying the printouts on my boards, they wanted to get their hands on the IBM keyboard and play my special-relativity game. I talked them through it while they typed in commands. "First you decide which effect you want to calculate: time dilation, length contraction, or mass increase. Then the program asks for the rest frame: duration, length, mass, plus speed of the object. Once you've entered those, it returns the relativistic time/length/mass as output."

I pointed to the lovingly hand-plotted graphs of the three effects of special relativity that I had pasted onto the triptych behind me. Then I explained that by typing commands on the keyboard they could add or subtract from any of the three factors to create different plots, like length versus speed, time dilation versus speed, or mass versus speed.

What really got their attention, though, was the fourth calculation I made by combining space *and* time to find invariant intervals, or spacetime. One of them actually shouted, "Wowee!" when INVARIANT INTERVAL flashed on the screen, just as I'd programmed it to do, the way a pinball machine flashes FREE GAME when you reach a winning score.

Then I recognized the two professors from USM who had first visited us at Heidelberg High. One of them stepped forward and shook my hand. "I hope you'll give Southern Miss a look when you're ready for college," he said. "You should come study physics with us. You have a huge head start."

AT THE END of the day, we broke down our exhibits and went out for a quick dinner. Since we hadn't eaten all day, we voted for the all-you-can-eat buffet at Shoney's. That took a while, what with the multiple trips to the buffet, and by the time we got back to the arena, the awards ceremony had already started. They had set up a podium on the gymnasium floor next to a trophy table.

The bleachers reserved for students and their families were full nearly to the top. As we climbed way up to the only empty seats, we passed the husband and wife who'd helped me earlier when my graphs blew away. They smiled and invited us to sit with their school group. We were happy to have a larger group to sit with, since most of the schools had a crowd of families and students. A couple of teams even brought cheerleading squads!

Before they got to the major awards, they presented a bunch of warm-up prizes from the military and industry sponsors. Each time a student's name was called, half a bleacher erupted in cheers and applause as the kid climbed down to accept the award. Some guy in an army uniform was going on about science and national defense, and then he called out my name: James Plummer Jr.! I didn't even hear what the award was for. I had to scramble all the way down the stairs to the gym floor to get my medal and then climb all the way back up to the nosebleed seats. When I got there, the white couple was clapping for me, along with my crew.

"Congratulations, James!" said the white woman. "You won an award. That's so wonderful." I was happy to have someone clapping for me, but she seemed surprised that I'd won.

After that, I won an industry-sponsor award for excellence. It was just a ribbon, and I had to climb all the way down the bleachers and then all the way back up. My legs were getting tired. But as I climbed back up, I saw the white couple clapping for me alongside Mr. Barber, Antrum, and Aristotle, which made me smile.

Then the major awards began. I won "Best Use of Mathematics in a Science Fair Project," from the American Council of Teachers, and this time, the white couple and all their students clapped for me too. Then came Biology, Chemistry, and a couple of other science categories.

When they announced Honorable Mention in Physics, and it wasn't me, I held the ribbons and medals in my hands as tight as I could. I started counting the iron beams in the roof rafters with my eyes, two at a time. When I got up to the sixteenth beam, the loudspeaker got all boomy and slowed down, as if the sound waves were traveling through water instead of through air: "And the First Place Prize for Physics goes to . . . James Plummer Jr. from Heidelberg High School!"

Antrum and Aristotle leapt to their feet and dapped me up and down, backward and forward. Even the white teachers tried to dap me, which ordinarily would have been awkward as hell, but I didn't care. I could barely feel my feet under me on my way down the stairs to the podium. I started counting the stairs as I went, partly to slow it all down and partly to keep from losing my balance. I counted the number of people in a row, and the number of rows in a single bleacher section, and the number of sections in the arena. Finally, I stopped counting and rode the applause the rest of the way down to the awards podium.

I'd won first place in Physics—and a big-ass trophy that proved I could bend spacetime just enough to make Mississippi stop and take notice of my beautiful Black mind.

33

ScIENCE FAIRS LET ME venture out of Heidelberg's little Black box and compete against white folks on their home turf. But Hattiesburg and Jackson were only a ninety-minute drive from home. My tuba took me a lot farther down the road, giving me a chance to explore the white universe outside my little neck of segregated backwoods Mississippi.

I'd made a lot of progress since that first week when Mr. Cross had me buzzing on my tuba mouthpiece. I learned the discipline of precision marching in formation, as well as musical notation and arrangement. By sophomore year at Heidelberg, Mr. Cross was letting me orchestrate and choreograph our football halftime shows. Mr. Cross's marching style and musical selection were super traditional, with no funk and no edge. I was ready to take things to the next level. I modeled our routines after the high-energy showmanship of all-Black marching bands and arranged the latest tunes from Black radio. I was always pestering Mr. Cross to help me find the most complex scores to orchestrate. When I looked at sheet music, the patterns of notes seemed to lift off the page. I found I could move the notes effortlessly with my eyes and repattern them to orchestrate different instruments.

The problem was that high school marching band was strictly local. We played during football games against neighboring teams and in three local parades every year. But each high school band director could send two musicians to State Band Clinic in Biloxi, which was all the way down on the coast. I was one of the rare students who got to attend the clinic all four years of high school.

Through the clinic I learned of other opportunities to compete outside of Heidelberg: summer band camp at USM in Hattiesburg was my first stop. There were only three Black students in a band of more than a hundred musicians. At the end of band camp, they held an awards banquet. I came in second place for the Students' Choice Award, which totally shocked me. It was basically a popularity contest, and I had no idea that people beyond my immediate friend group liked me. I had always been a leader in the Heidelberg High marching band. But Students' Choice in a white cracker band in southern Mississippi? That was a puzzle.

The band directors selected the two top merit award winners, one from each of the two bands. Even though I was first tuba, second band, I was stunned to win the Director's Award for second band. I knew I could compete as a musician. But what threw me was the idea that white band directors would choose a Black kid for a merit award when there were so many white students to pick from. I just assumed they'd be prejudiced. It felt different from winning first prize at the state science fair. I expected science professors to judge projects on their merit. Music felt more personal. More racial, somehow.

Encouraged by my success at band camp, I auditioned junior year for the most elite high school band in the state: the Mississippi Lions Club All-State Band. Mr. Cross liked to tell us about the Heidelberg trumpet player who made the Lions band back in the early 1970s. He went on to have a jazz career and become the assistant band director at Jackson State University. Though several local students auditioned in the years that followed, no one had been selected since.

I aced my audition and made the Lions Club All-State Band. That was the good news. On the downside, I had to spend two weeks that summer rehearsing in Mississippi heat and humidity in advance of the nationwide competition in San Francisco: the Lions Club International Parade of Nations. The heat was brutal. Our feet blistered. Kids passed out from sunstroke and dehydration. The tuba never felt heavier and the sun never felt hotter. The Black members of the band—just seven of us among the two hundred musicians—felt it was a point of honor not to puke or pass out on the field.

Since Lions Club was a nearly all-white Mississippi band, our program songs and chants were filled with shout-outs to the Old South and the Confederacy. Our farewell concert—performed for our families before we set off on the cross-country bus ride to San Francisco—was dedicated to "Our gallant boys in gray," and our finale number was "I Wish I Was in Dixie Land." When it came time to sing "Oh, I wish I was in the land of cotton/Old times there are not forgotten . . ." most of the Black band members simply lip-synched the lyrics. I kept my mouth fastened on my tuba's mouthpiece.

When we got to San Francisco, everyone made a big fuss about the trolleys, the Golden Gate Bridge, and our band winning first prize at the competition. But for me, the most memorable moment happened while we were waiting in line to play at Fisherman's Wharf. I noticed that the band behind us had red maple leaves on the flags at the front of their formation.

"They must be from Canada!" I said to the one Black snare-drum player in our band. "Let's go talk to 'em." I'd never met a Canadian before and was curious as hell to hear what they sounded like.

We slipped out of formation and started to mingle with the Canadians, as inconspicuously as two Black dudes carrying a sousaphone and a snare drum could manage. We carried our own social rules along with us. Unbreakable Rule Number 1 was: Never approach a white girl. Even after participating in so many statewide band events throughout high school, I had never once held a conversation with a white girl. White dudes, I could talk to all I wanted, and most of them were down with me. But since I'd returned to Mississippi in seventh grade, I'd had exactly zero interactions with white girls. That's how you got yourself killed in my neck of the woods.

We tested the waters by chatting up the drum major, a tall guy with a taller hat from Toronto. Then two white girls with piccolos sauntered up to us and said, "Hi," in unison with synchronized little wrist waves. "Pretty boring, just standing around like statues all morning, eh?" said one of them.

She said this like chatting up two Black dudes was something she did all the time. Faced with my first conversation ever with a white girl, I was too shocked and suspicious to speak. Smiling white girls luring you to your death was exactly the scenario I'd been warned against growing up.

Finally, the snare drummer started in talking with them, and pretty soon the four of us were chatting back and forth like we'd grown up in some interracial TV sitcom.

A few minutes later when we fell back into formation with our band, the other Black guys started peppering us with questions. "What did they say?" "What did *you* say?" "Did you get their numbers?"

"Canadians are cool!" was all I could respond. I couldn't remember a word anyone had said. I just knew my mind had been blown.

34

IF YOU STRETCH IT out straight, a B-flat tuba has eighteen feet of tubing. By the end of junior year, those eighteen feet were starting to look like my bridge to college. My success with the Lions Club band led to a spot on the All-South Honors Band and a nomination to the McDonald's All-American Band, which only accepted two musicians from each state and played big national events like the Macy's Thanksgiving Day Parade and the Tournament of Roses Parade.

I started getting letters and brochures from most of the top Black college bands I'd grown up watching on TV: Florida A&M, Alcorn State, Grambling State, Tennessee State, Alabama State, and the band I dreamed of marching with: Jackson State's Sonic Boom of the South, which fielded thirty tubas!

I even got letters from schools I'd never heard of, like Northwestern University outside of Chicago. When I looked at Northwestern's brochures with pictures of their orchestra program, I started thinking beyond marching bands. I imagined becoming a professional tubist and playing in an orchestra after college. I needed to talk to Mr. Cross about how all that scholarship business worked. Maybe I could get one to Northwestern.

But that's not how it played out for me.

One morning during homeroom, a voice boomed across the school PA system: "James Plummer, please report to the principal's office immediately. Repeat . . . James Plummer, please report immediately to the principal's office."

Everyone in class turned to look at me. I figured I'd been caught out for something bad I'd done, or had been accused of doing. I wonder who done lied on me now, I thought as I walked to the principal's office.

Principal Green was waiting in his outer office when I arrived—six feet three inches of clean-shaven, slick-haired, suit-wearing, upright, dead-serious Mr. Leader-of-the-School. I stood up as straight as I could so I wouldn't have to crane my neck to make eye contact while he chewed me out. He didn't invite me to sit down or anything. I guess he liked the height advantage.

"What are you doing after high school, son?"

That wasn't the question I expected. "I'm planning on going to college," I said. I didn't really have a plan, but I'd scored higher on the ACT than anyone ever had at Heidelberg High. And my grade-point average had me on track to become class valedictorian.

"College? Really?" he replied. "How you gonna pay for that?"

The truth was, I didn't have a clue how money and college worked. I didn't have any friends or family who'd gone to college. Neither of my parents knew anything about college, and they'd never talked to me about my future education, much less how to pay for it. The only kids in my class who were planning to go to college had parents who were teachers at Heidelberg. But I didn't want to tell Mr. Green all that. So I mumbled something about recruitment letters from colleges with marching bands and possible scholarship opportunities.

"So playing in a marching band's a paying profession now?" Mr. Green scoffed. "Come with me, son. I want to introduce you to someone."

He led me through the door to his private office. Inside stood a tall white man—taller than Mr. Green, even—dressed sharp in a navy officer's dress-white uniform. He smiled and extended his hand toward me.

"I've been wanting to meet you, James Plummer Jr.," he said. "I'm Senior Chief Gauge." He sounded like he was truly proud to meet me, like he was a coach who'd found a star quarterback at a small-town school where no one had thought to look. "Young man," he said after he'd released my hand, his eyes still fixed on mine, "you got the highest score on the ASVAB that has ever come across my desk. You did well in every single category of questions. I've never seen anything like it."

A few months earlier, all the students at Heidelberg High had been ushered into the lunchroom to take the Armed Services Vocational Aptitude Battery, or ASVAB. As usual, I had been the first one to finish the test—and had promptly forgotten about it.

"Look, the most I can offer you is $20,000 a year," said Senior Chief Gauge. "But if you go nuke—and I know you will—then after two years, if you reenlist, we promote you and give you a $30,000 signing bonus. How's that sound?"

My mind was struggling to comprehend what he was saying. Mama once bragged that Dwane had earned $15,000 in one year. This guy just offered me $20,000—and then two years later, $30,000!

Mr. Green spoke up. "What do you think, son? It's a terrific opportunity he's offering you."

"What's 'going nuke' mean?" I asked.

"Well, you'd be a nuclear engineer," said Senior Chief Gauge. "The navy runs more nuclear-powered submarines than every country on Earth put together, and we've never had an accident. We'll teach you how to be a nuclear engineer and if you want to go to college, we have ways to help you do that too."

I paused another moment to let it all sink in. "Y'all gon' pay me $20,000, you say?"

"You don't start at $20,000," Gauge responded. "You start as a navy recruit in boot camp. But if you get into the Nuclear Field Program, after boot camp you go to nuke school in Goose Creek, South Carolina." Gauge smiled down at me.

"My goodness!" Mr. Green spoke up. "That sounds like quite an opportunity. So what do you say, son?"

I didn't know what to say. I couldn't believe someone was offering to pay me twenty, maybe thirty thousand dollars to become a nuclear engineer. The Reagan recession had hit our community hard, and everybody in my family was out of work. Howard Industries laid off half of its factory workers, landing Dwane and Mama on the relief rolls. Dwane hurt his back working on the assembly line, so he couldn't haul pulpwood, which was the fallback cash-flow solution for many men in our part of Mississippi. Even Bridgette, who'd been working fast food at Taco John, got laid off. Just the week before, Dwane and Bridgette had rented out their trailer and moved into the house with Mama and me. Daddy had retired with a pension from Kaiser Aluminum by then, but he had a new family and wasn't providing for any of us.

The only work on offer—and only for women—was cleaning white folks' houses. Both sides of my family, the Plummers and the Alexanders, were proud people, and cleaning white folks' houses was the lowest rung on the employment ladder. Mama refused to stoop that low, but Bridgette was cleaning up after white folks most days now. I picked up some household plumbing work around Piney Woods and Heidelberg, and in the spring I filled out folks' tax forms, charging anywhere from $10 to $20 to get them some money back from their withholding. And as a side hustle, I had my own weed grow. Every Friday morning I sold a dozen joints at school for a dollar apiece—but that was just for drink money at the clubs on Saturday night.

All of that didn't amount to pocket change compared to what the navy was offering me. And hadn't Senior Chief Gauge said something about how they might help send me to college?

In my conservative and patriotic Mississippi community, military service was held in high esteem. Every year our school hosted Military Appreciation Day, which included essay and speech contests. In eighth grade I'd won a $10 gift certificate to the Dairy Queen in

Meridian for a speech I gave called "Why We Need a Strong Military to Keep America Safe." My daddy had served in the army, and I grew up knowing a lot of Vietnam vets who were respected for their service. Every year several guys from Heidelberg joined the military immediately following high school. Some of them ended up in Germany or Guam or some other faraway-sounding place. I'd never seen myself going in that direction, but then nobody had ever offered me cash money and the possibility of a college education before.

Mr. Green and Senior Chief Gauge stood in front of me with expectant smiles on their faces. Finally, I said, "I gotta talk to my mama and see what she says. But it does sound like a good opportunity."

I WAITED TILL after dinner before telling Mama how much money Senior Chief Gauge had offered me to enlist, and how I needed parental consent since I was a minor. Before I even got to tell her the part about going nuke, she picked up the pen from on top of her crossword and said, "Gimme what I gotta sign. Hurry up, 'fore the navy comes to it senses."

Mama printed out and signed her name: Elaine Josephine Alexander.

"I guess my Li'l Jame a bonus baby now," she said, shaking her head in disbelief.

Senior Chief Gauge had already filled in my name above the enlistee's signature line: James Edward Plummer Jr. right next to the navy logo of an eagle holding an anchor in its claws. I took a deep breath, counted the six red stripes on the eagle's breastplate, and signed.

I was in the navy now.

35

THE BEST THING I got out of the navy was the encouragement Senior Chief Gauge gave me before I even got to boot camp. I'd never met an adult who was as enthusiastic about me as Senior Chief Gauge was. Some teachers supported me along the way, and Mr. Cross taught me the value of discipline. But Senior Chief Gauge was different. He treated me as if I were his golden find. He also complimented me frequently on my "high character." No one had ever done that. Not any of my teachers. Not any of my family members. Not anyone. Now here was this navy officer treating me as if I was smart *and* good.

Right away, Senior Chief Gauge was focused on getting me into the best possible programs. The first order of business was for me to take the Navy Advance Programs Test, or NAPT, which I had to pass to qualify for the Nuclear Field Program. Senior Chief not only got me scheduled to take the exam over in Jackson, he drove me there himself.

All the way up to Jackson he talked about what my life was going to be like in the navy. "Man, I know you're gonna go nuke!" Senior Chief Gauge said. "These guys up here are going to be blown away once they meet you and see what you can do."

I never suffered from a lack of confidence in my abilities. But hearing Senior Chief talk about me made me believe in myself even more. And it made me believe in him too.

All that confidence-boosting was like rocket fuel for me. I aced the NAPT, getting the highest score ever for the Jackson district. Next, Senior Chief Gauge got me into a program called BOOST, which stood for Broadened Opportunity for Officer Selection and Training.

"BOOST is for turning enlisted men into officers," he explained. "It's focused on people like you who come from the country or the inner cities where the education system isn't as good as in the sub-urbs." He explained that I'd go in as an enlisted man and they'd give me a year of military and academic training. After that, I could get a Navy ROTC scholarship to attend college.

Senior Chief Gauge had plotted a course for me that would get me an education, a career, and a path out of the Mississippi woods and around the world. I was on my way!

Or so I believed.

HERE'S THE HIGHLIGHT and lowlight reel of my brief career in the U.S. Navy:

Boot camp outside of San Diego was pretty much like you've seen in the movies. Head shaved. Piss-in-a-cup drug-tested. Indi-viduality shouted and beaten out of you. Saving grace: I got to play the contrabass bugle almost daily at commissioning and graduation ceremonies. Best takeaway: I learned to keep my clothes and gear clean and tidy.

Another bonus: the navy taught me algebra. The whole point of the BOOST program was to make up in one year for the twelve years of crappy public school education. Despite everything I'd taught myself, I never learned algebra until the math classes in BOOST.

The navy also taught me what institutional racism looked like from the inside. Four hundred and fifty of us began the program. They'd whittled us down to two hundred by spring—and almost all the ones who got bounced out were Black or Brown. We also pulled all the worst work details and got shafted when it came to merit awards.

I was doing just fine. Top of my class academically and headed straight for nuke school. Then my skin flared up, and I was diagnosed with atopic dermatitis, which somehow hadn't shown up during my induction physical. It turned out that having atopic dermatitis disqualified me from serving on ships.

The good news was I finally got a diagnosis for the rashes that had afflicted me since I first came to Mississippi, and they gave me some prescription creams that helped tamp it down some. The bad news: they offered me an honorable discharge and a flight back home. So much for my fantasy of becoming a nuclear engineer.

A week later, I landed with a thud back in Piney Woods, without a plan B.

36

FELT MORE CONFUSED than angry about my sudden discharge from the navy. Here's the thing about enlisting in the military: they take away all your personal stuff and as much of your identity as they can—your clothes, your hair, your first name. In exchange, they give you everything you need to survive—which for me at the time was an okay deal. They give you food, a uniform to wear, and a place to sleep at night. Most of all, they tell you exactly where you have to be and what you have to do every waking minute of your day. You stop having to take charge of your life.

I got through all the humiliating and dehumanizing shit they laid on you in basic training by fantasizing about my future. I'd lay in my bunk before going to sleep and imagine myself in an officer's uniform as chief engineer aboard an Ohio Class nuclear submarine. Being air-dropped back into Piney Woods was like waking up from a dream of ocean-faring adventuring and finding myself fishing with a bamboo pole in Kelly Hill Pond again. I was an unemployed nineteen-year-old with no plan or vision for my future.

I was shocked to discover—since no one had told me—that everyone had moved out of our house in Piney Woods. Mama couldn't get work and couldn't make the house payments, or even the payment on the furniture, which was all on installment. When a truck came to repossess the furniture, Mama had moved back to New Orleans, and Bridgette and Dwane had moved back into their trailer. They invited me to live with them, along with their two-year-old daughter.

Ending up back with Bridgette and Dwane in their trailer felt like a huge step in the wrong direction. But I had nowhere else to go.

THE OTHER THING that got me through the toughest parts of boot camp and BOOST was thinking about my hometown girlfriend, Lisa. So I naturally turned to her for solace when I arrived back in Mississippi with no real home or future plans. Lisa was sort of family, being Dwane's little sister, the tenth of fourteen Morgan kids. Growing up, I'd always thought of Lisa as a pesky little girl—until she showed up at Heidelberg High as a freshman during my senior year, looking fine and shapely with beautiful long, thick hair. Suddenly, all my buddies at school were asking me to introduce them. I started to take notice of Lisa and decided to introduce myself instead.

I was smitten by Lisa's beauty, innocence, and sweetness. In fact, she went by the nickname "Sweet" in high school. It seems sort of old-fashioned, looking back, but I fixed on Lisa as the image of female beauty and purity. She was a virgin, and so was I, if you didn't count a couple of hurried encounters with a majorette in the back of the band bus. Lisa touched something deep inside me. I guess it was the innocence that had been snatched away from me when I was a kid, and I had pretty much given up on retrieving.

When it came to courting, my community was very traditional. Every Sunday, the boy would show up at the girl's house and hang out with her on the couch at a decent distance while some older family member sat in the room with them. After some Sunday visits, the boy might gain the trust of the family and the young couple would have some time alone. The boy would spend holidays at the girl's house with her family. Then the two of them would visit his family. There would perhaps be dates when the girl was seventeen or eighteen. And once both of them had graduated and at least one was gainfully employed, they'd marry.

I imagined that would be my path with Lisa. When I went off to the navy, I assumed we'd marry when she graduated high school and I'd graduated from nuke school. I wrote her letters from boot camp and from BOOST, and she sent me a sweet picture of herself to tape up over my bunk so I could keep her in my head, and in my heart.

When I came home from BOOST training for Christmas, I gave her a promise ring, and we had sex for the first time. Probably because we were both newbies at that activity, and pretty inept, our romance still felt innocent and pure—which is how I wanted it with the woman I saw as my future wife. Lisa seemed like a princess to me, and I hoped to be her prince in a navy officer's dress whites.

Now that I was out of the navy and back in Heidelberg, I had to rewrite the script. No adults I knew had any advice. As far as my parents were concerned, you were either doing something, or doing nothing. Any job counted as "something." There wasn't any discussion about careers. I'd never met a lawyer and before enlisting in the navy the only doctor I'd ever seen was during my physical for seventh-grade football.

Everyone I knew in Mississippi was just doing whatever they could to get by and survive. Jobs were hard to find. Mississippi unemployment for Blacks, which hovered around 15 percent during most of the 1980s, reached its peak in 1986, the year I left the navy. I put in applications at every store and fast-food joint in Laurel, but to no avail. I was a regular at the unemployment office. They had a special line for military veterans, which I thought might make it easier for me to get a job. No dice.

So I lived off my unemployment checks and had absolutely no hope or vision for my future. My life centered on being with Lisa as much as possible during the days and hanging out with my boys at night.

My two closest friends in Wesley Chapel were Chris Morgan, who was one of Dwane's younger brothers, and Johnny Garfield, who was from Chicago—what everyone referred to as "Up South"—which made him seem more sophisticated and worldly. Johnny Garfield wanted us all to call him JG, which he thought sounded cooler. I'd known JG since we were twelve-year-olds at AME Church and smoking weed in the woods together on weekends in high school. He graduated high school a year ahead of me and headed off to Tougaloo College outside Jackson. As soon as he came home from school that summer, he was on me to join him at Tougaloo, saying, "You're too smart of a dude to just get a local job and work in a store or factory the rest of your life."

I'd never heard of Tougaloo before JG went there. Everyone in Heidelberg knew the big-name Historically Black Colleges and Universities, or HBCUs, in Mississippi—Jackson State, Alcorn State, and Mississippi Valley State. But Tougaloo?

"Do they have a marching band?" I asked.

"No. And no football team neither. They do have about five times more girls than boys—but that's an opportunity, not a problem. The coeds are friendly. They ain't like those stuck-up girls at Jackson State and Alcorn."

I'd been home for six weeks by then and still hadn't found a job. College—even a small school with no marching band—was starting to look like a better option than living in a trailer with my sister and her family.

I called up Tougaloo and found out that I was prequalified for admission because of my grades in high school and my ACT scores. And since I graduated at the top of my class, I was eligible for a scholarship. But that deadline had passed. Over the next month I was able to patch together a Pell Grant, a scholarship from the Teagle Foundation, and a student loan that covered my tuition. Tougaloo said they'd find me a job once I was on campus.

WHEN IT CAME TIME to head off to school, I reminded Lisa that Tougaloo was only a few hours away and I promised to visit. I wasn't worried. We'd said a tougher goodbye when I went off to the navy, and we'd hung together as a couple through that.

Next, I drove down to the Goose in New Orleans to say goodbye to my parents.

"You wanna smoke a jernt?" Daddy asked as soon as I'd walked through his door. His country accent sounded stronger to me after a year away. We sat at his kitchen table, which was piled high with a new shipment of weed, and he watched me hitting the joint. I hadn't smoked the whole time I was in the navy, but I'd been making up for lost time since my discharge.

"Look, I see you a grown man now," he told me. "You're gonna do what you wanna do. The streets are bad, it's no joke. You wan' some weed, you wan' some blow, you wan' pills—you come home for it. I don't wan' you in the streets. Whatever you wan', you git it from me."

I drove across town to see Mama over on America Street, where she was living in the house of her recently deceased stepmom. Mama had a serious demeanor when she came out onto the front porch to say goodbye. Even though Tougaloo was in a sleepy little suburb of Jackson, Mama didn't like what she was hearing about goings-on in the state capital. Drive-by shootings, AIDS, crack. She gave me a long hug.

When she pulled away, I saw she had her white pearl-handled .22 in her hand. The same little handgun she'd lent me to take to tough clubs in Laurel when I was in high school. Now, she pushed the gun into my hand, closed my fingers around it, and repeated the same warning she'd given me at age fifteen when I first went off to the Troubadour Club. "Just don't take it out unless you gonna use it," she murmured, turning away so I wouldn't see her start to cry.

PART THREE

HISTORICALLY BLACK IN COLLEGE

The darker the night, the brighter the stars.

— Apollon Maykov, 1878

37

Tougaloo was known as a "cerebral" HBCU with a reputation for educating future doctors and lawyers. Since it was too small and too female to field sports teams—which were everything at most HBCUs—academics dominated campus life. The student body was all Black, and the humanities professors were mostly Black. But with so few Black PhDs in the sciences in the 1980s, there wasn't one full-time Black professor on the sciences faculty.

My first precalculus class was an eye-opener for me. The professor, a balding Egyptian named Dr. Nimr Rezk, lectured at us while writing math expressions on the blackboard, with only occasional glances over his shoulder in our direction. The students all sat silently in their seats, writing in their notebooks. I looked around the lecture hall thinking, What the hell is going on? I didn't hear him give us an assignment. How do they know what to write and I don't? The whole idea of lecturing and note-taking was new to me. On day one, I already felt outclassed and unprepared.

Honors English was taught by Dr. Jerry Ward, the pride of the Tougaloo English department. A light-skinned Black man, Mississippi native, and Tougaloo alum, Dr. Ward was renowned as a poet, essayist, literary critic, and Richard Wright scholar. He informed us on the first day of class that as honors students, we faced higher expectations. We needed to participate in class discussions and our written essays would be held to the highest standards.

When Dr. Ward gave us an essay topic to write about in class the first day, I banged out a formulaic page of prose. I was the first to finish, per usual. As I sat in the back row waiting for the others to catch up to me, I looked at what the student sitting next to me was writing. His vocabulary and compositional style were way more advanced than mine. And he was a weed-smoking pal from my dorm! If he wrote circles around me, what were all the square kids writing? When Dr. Ward selected a student to read her essay aloud, the complexity of her sentence structure made me think, Maybe I'm in the wrong class. Or school.

Ever since high school I made a habit of reading the textbook cover to cover the first week of class. I did the same at college. But when I went to lectures, I was confused by how the professors explained the material. In high school I often knew more about the subject than the teachers. College professors were in a different league. And they didn't just teach from the textbook. They lectured. No matter how hard I listened, there seemed to be something blocking the space between my ears and my brain. College was the first time I realized that I learned differently from the other kids who sat quietly with notebooks open and pencils at the ready. The whole lecture-hall scene made me so restless and anxious that I had to sit in the back row with one eye on the exit.

Once I figured out that attendance was optional in college, I stopped going to lectures. I'd show up on important dates like the first day, last day, and just before exams. But that wasn't nearly enough to get good grades. My education to date had been measured against very low expectations. Being smart and competitive was all I'd needed to succeed at Southside Middle and Heidelberg High. The material I had to master at Tougaloo was much more complex, and the expectations for achievement much loftier.

As my academic performance declined, so did my self-confidence. Ever since primary school, I'd been academically dominant. It was a blow to have to face up to my undereducation and unpreparedness for college-level study.

THE SOCIAL FRONT wasn't any more inviting than the lecture hall. I'd never met Black kids like the students at Tougaloo—raised up inside middle-class families with educated parents. They were polished. I was crude. A lot of them looked at me like I was some kind of primitive species of backwoods Black man, which I basically was, compared to them.

Even though there were many more girls than boys at Tougaloo, freshman guys were low men on the status pole. The sorority sisters were looking for boyfriends who had PEP—Potential Earning Power—which I clearly didn't. I felt lucky to have a hometown girlfriend to visit in Heidelberg on holidays and long weekends. It was a relief to be somewhere for a few days where I didn't have to pretend I wasn't country ghetto.

Since we weren't allowed to pledge fraternities until sophomore year, we freshmen had to form our own social groupings. I naturally gravitated toward the dudes I felt comfortable with: my weed-smoking buddies in the freshman dorm. Most evenings you could find us in one of our dorm rooms smoking weed together and listening to Steel Pulse or Black Uhuru on a high-powered stereo.

I DON'T KNOW WHY I told JG about my daddy's offer to be my plug and supply me with whatever drugs I wanted. Maybe because I was hanging out with stoners where weed was the main social currency. When JG found out I also had a gun, he got all amped up about the idea of us becoming Tougaloo's premier weed dealers. He wouldn't let it rest until I made the call.

"Hey, Daddy!" I said into the hall phone. "It's James Plummer Jr."

"Hey there, Li'l Jame! How you doin'?"

"I'm thinking about coming down there to visit you this weekend. What you think?"

"I think I'll be happy to see ya," he said, laughing in that warm way he had. "Come on down."

"One more thing. Remember our talk this summer, when you told me to come see you if I needed anything?"

"I do."

"Well, I'ma take you up on that offer."

"That all right too."

"I'ma bring my friend JG with me. I've known him since seventh grade in Heidelberg. He cool."

"That's cool with me, then. Bring him along."

38

JG AND I drove the three hours down I-55, arriving at Daddy's house in New Orleans East by early evening. He greeted me at the front door with a handshake and a one-armed hug.

"Come on in! Good to see y'all." I introduced JG. "Nice to meet ya," Daddy said, extending his hand.

"Good meetin' ya, Mr. Plummer."

"Come on in and let's go to the kitchen."

My older half-brother Byron was sitting on a stool at the bar that separated the kitchen from the dining room. I hadn't seen him in over a year. "Byron! I didn't expect to see you." I dapped him up and we gave each other a one-armed hug.

"Daddy told me you was coming. How you been? I thought you was in the navy."

"They kicked me out." I shrugged. "I just started college. Up at Tougaloo."

"That probably what you was s'posed to do anyway," he said. "They always say you some kinda genius."

"Nerd might be more like," I said.

"Nerd?" he guffawed. "Boy, you James Plummer Jr.! You more than just a nerd, nah."

"Well, my friends say I'm a cool nerd," I said.

"Well, then, stay cool, and stay in dem books."

For his part, JG just remained quiet and peeped the scene. That's standard hood etiquette—keep your mouth shut when you don't know anyone.

Daddy came over to the table with beers for JG and me. "Y'all wanna smoke a jernt?"

"Yes, sir," I replied with a big smile.

"Byron, fire up one dem jernts," Daddy said. Byron lit a joint from a pile already rolled up on the bar. He took a long drag and passed it to me.

Meanwhile, Daddy picked up a clear glass pipe from the bar, held the flame of a small blowtorch to the end, and hit it. Then he passed it to Byron. He hit it just as Daddy had done and exhaled a cloud of white vapor that didn't have much of a scent. Byron held on to the pipe, then he looked at Daddy and motioned his head toward JG and me. Daddy nodded.

"Y'all want a hit?" Byron asked.

"What is it?" I said.

"It's rock. Rock cocaine."

I looked at JG, then answered, "Naw, I'm good."

JG rose from his seat, all excited. "I'll hit it," he said.

I'd grown up around weed and had been smoking it almost daily since I was thirteen. But I wasn't much of a drinker, and I wasn't interested in harder drugs. I'd avoided speed all through high school, and I made sure to steer clear of angel dust, which scared me silly. I'd been in enough juke joints and clubs to see how dudes behaved after they smoked dusted joints. Like zombies whose souls had left their bodies in a swirl of smoke. By comparison, coke seemed pretty tame. I'd seen folks taking bumps of powdered coke off their car keys at parties over the years, including my mama, Daddy, and Andrea. Rocks were new, but they didn't seem like a big deal.

JG took the pipe and put it to his lips while Byron worked the torch for him. JG inhaled the vapor, held it for a long time, and then exhaled. A second later he uttered a small moan and inhaled again deeply as if preparing to sigh.

"How you like it?" he asked JG.

"That's goooood."

"You sure you don't want none?" Byron asked me.

"Naw, I'm good."

"You smoked before?" Daddy asked JG.

"No, sir. I've seen folks smoke up in Chicago but this my first time."

"It's a clean high," Byron said. "The best high there is." He hit the pipe again.

I looked at them and for the life of me couldn't see any of the usual signs of a high person. Their eyes weren't red. They didn't slur their speech. But then I noticed something fierce in the way JG was looking at the pipe. Like he couldn't wait to hit on it again.

"Byron," Daddy said, peeling a couple of bills off a wad, "take JG with you up to We Never Close and get some shrimp and crawfish. Buy some mo' beer too."

After JG and Byron left, Daddy sat down across from me at the table. "So whatchoo trying to do?"

"Well, I'm in school and I ain't got no job so I thought I'd do a little business. There ain't nobody selling on campus. And the stuff in town ain't that good. There's some good 'sess here and there," I said, referring to a seedless strain of weed, "but mostly they got that ac-a-dem-ic stress."

"How much you thinkin' 'bout movin'?"

"I was thinking a couple ounces or something."

"A couple ounces?" Daddy laughed. "Man, if you the only one in town with good weed, that ain't gon' last you long at all. I'll give you a couple QPs." Quarter pounds were the smallest denomination my daddy was dealing in those days.

"How much you want for it?"

"You ain't gotta pay me. Not till you get your business feet on the ground."

39

As soon as JG and I made it back up to Tougaloo, we headed straight for the stationery store to buy a box of small yellow envelopes. We decided to sell $20 "bags," since that's all a typical college student could come up with.

I went over to JG's room the next day and found that he'd acquired a triple-beam balance scale from who-knows-where. It looked like the kind we used in the general chemistry laboratory. We spent the evening smoking in JG's room, listening to music, and cleaning and sorting the weed into envelopes. JG decided that he'd sell $1 joints as well, so he rolled a fistful of those. I wasn't going back to dealing joints like I was still in high school. Selling the smaller quantities was more profitable but meant more customers, and more risk. I wasn't worried about getting busted so much as blowing my image as a smart kid and aspiring science nerd.

I set aside half my quarter pound to smoke with my dorm buddies, and divvied up the other half into fifty-six $20 envelopes. That would net me over $1,000! I wondered if I could sell them all.

In true JG fashion, he told everyone he knew on campus that we were in business. When word spread of the quality of our product, it started to move. I was concerned about security, so I stopped dealing at eight o'clock each night. If someone came looking for weed who I didn't recognize, I'd hand him a joint and say, "Fire it up, and see if you like it." I figured an undercover cop wouldn't do it. I also resolved that I'd sell no more than $100 worth a day.

After three weeks we were completely sold out. I laid low and spent my money on gas, books, food, and beer. JG was flashing wads of cash all over the place. He'd take a Franklin to the bank and change it for a hundred $1 bills. Then he'd wrap a few hundreds around the ones, just for flash. Lucky for him we were on campus and not in the hood. On the street, he'd have been robbed in a blink.

Two weeks later, we went back down to New Orleans to make another pickup. This time, JG and I each kicked in $300. It went pretty much like the first time. Daddy was casually smoking rocks. JG joined in, and I didn't. Daddy told us that coke was flooding in from everywhere, but his weed connections were drying up. After that, we could only score one QP at a time.

At the end of the fall semester I got my first report card and was stunned at how poorly I'd performed. When I first got to Tougaloo, I had no idea how to choose a major. I thought to myself, Well I'm a smart science guy, I'll be pre-med. But the thing I didn't understand about college is that unlike high school, being smart and having a reputation for smarts means nothing. Getting good grades required focus and hard work—and I brought neither to the table. Cutting labs turned out to be an even worse strategy than skipping lectures. My second semester grades pretty much blew up my self-image as a doctor in training. C's didn't get you into medical school.

AS I ROUNDED THE CORNER into sophomore year, I clung to the hope that pledging a fraternity would turn things around for me. In the absence of a football team and marching band, fraternities and sororities were the center of the social universe at Tougaloo. There were four traditional Black fraternities on campus, each with its own rep. Omegas were the boisterous jocks. Alphas were more academically inclined. Sigmas were thought of as country. I wanted to pledge Kappa Alpha Psi because their motto was "Achievement in Every Field of Human Endeavor." I could get with that. And I embraced the trial by fire of pledging for an elite and hard-to-get fraternity.

But first I had to endure a brutal hazing of paddling, sleep deprivation, and verbal and physical abuse honed over decades to push you to your limit. Pledging Kappa Alpha Psi was tougher than navy boot camp, though it was based on the same principle: You don't know how strong a rope is till you break it. Once you broke—and everyone broke eventually—you came together as brothers.

To outsiders, pledging could appear ridiculously, even dangerously, abusive. But it was abuse with a purpose. As the big brothers explained to us: if we could survive pledge, we could thrive in the world as brothers. Hazing was framed for us as a metaphor for the trial that America's race-based social hierarchy had in store for us once we graduated into adult life. Our brothers were giving us impossible tasks to perform and placing us under extreme pressures for a few weeks to enable us to become part of an Achievement Brotherhood of Black Men, for life. I was prepared to endure anything for that.

GETTING A CHANCE to connect with Jessica was a perk of pledge season I never saw coming. She was out of my league. I was country ghetto, she was country club. I was a stoner nerd from the backwoods, and Jessica was Tougaloo royalty. Literally. The Homecoming Court elected her to represent their class as Miss Sophomore and Miss Junior—an honor that girls actually campaigned for. Jessica won both years as a write-in candidate! She was beautiful, with a dignified and poised bearing. She grew up with two parents in one of the better neighborhoods of Natchez, drove a nice SUV around campus, and belonged to America's first Black sorority, Alpha Kappa Alpha. Their motto was "By Culture and Merit." And just as the men of Kappa Alpha Psi were nicknamed "The Pretty Boys," the ladies of AKA called themselves "The Pretty Girls."

I didn't know any of that when I first encountered Jessica in chemistry lab. General Chemistry had been my strongest subject freshman year, and my professor, who knew I was desperate for a paying job, hired me as his teaching assistant. Jessica and her friend Veronica were both in the section I taught. I couldn't tell whether Jessica was hitting on me or if she kept calling me over for explanations because she didn't understand polymers. Finally, after the fourth time she waved me over for help, Veronica blurted out, "Girl! Can you stop flirting so we can get some work done?"

Jessica chuckled and dismissed me with a sly smile. "That's all for now. Thank you."

Over lunch with JG, I said, "I think this chick in my chemistry lab likes me."

"Yeah? Who dat?"

I told him.

"Awww, Plummer, man. You can't fuck with Jessica. That's Denard's woman."

"No shit?!"

I knew who Denard was. Everyone did. He was the star shooting guard on the basketball team. Since basketball was the only real sports team at Tougaloo, that made Denard a big man on campus.

"Yeah, man. Forget about her."

"Damn! She fine as hell too! You sure? 'Cause she the one flirting with *me*."

"Maybe she is and maybe she ain't. But they been together like forever. I think they started dating in high school. Forget about her. There's hella girls around here anyway. As a matter of fact,

Trudi was asking me about you the other day. You know she got some ass!"

"Maaaannnn, sheeiiiit! I'm scared of Trudi. She got too much ass!"

We both busted out laughing.

40

DURING PLEDGE PERIOD we'd have to sleep in our clothes because our big brothers could, and would, wake us in the middle of the night to make us recite frat history while standing on one foot while several others yelled at us, or gave us some impossible task to perform. Failure at any task resulted in physical punishment for either you or your line brothers.

One day our big brothers informed us pledgees that Kappa Alpha Psi brothers from all over the state would be converging on Jackson. They warned us to find a place to hide out for the night because the visiting Kappa brothers might wreak some serious havoc on our bones.

My line brother Hannis Longino was planning to hide out with a girl named Carla, who lived in the same dorm as Jessica. Shortly before midnight, we arrived at the ground-floor lobby, which was as far as boys were allowed to proceed in the all-girl dorms. Carla came down to meet us and, with help from a lookout, snuck us upstairs to the third-floor hallway. Hannis tried to get me to knock on Jessica's door, but I was too chicken. So Hannis rapped on her door and ran away down the hall with Carla.

A moment later the door opened, and Jessica was standing there in her pajamas.

I didn't know where to put my eyes, or what to say. I looked down at her ankles and whispered, "A bunch of Kappa brothers are coming on campus tonight, and I have to hide out." I raised my gaze up to her eyes. "Can I stay in your room?"

"Yeah, you can," she responded in that sweet Southern belle voice of hers.

At Tougaloo everyone took pity on the pledgees and tried to help us out. Sometimes this meant giving us food. Sometimes it meant helping with our academic work. And sometimes it meant allowing us to hide out in their dorm rooms. So I figured Jessica was just taking pity on me. As I followed her back into her darkened room, I could tell that her roommate, Kenya, was in the twin bed across from hers.

Since I hadn't showered in a week—what with all the chaos and mayhem of pledge period—I thought it polite and respectful to lie down on the floor between the twin beds.

After a few moments, I heard Jessica's voice calling softly in the dark. "Come up here and sleep in the bed with me." My brain tried to uncover the logic beneath that invitation. Jessica was Denard's woman. Jessica was out of my league. I was a dirty, smelly mess. What was wrong with this picture?

I peeled off my shoes and socks and crawled into the bed with her. Then, for the first time since I was five and sharing a bed with Bridgette, I cuddled with someone. Jessica laid her arm across my midsection and scooted up close to me. Her cheek was touching mine. I'd never experienced this level of intimacy or the gentleness of a female touch in this way. I'd never felt such acceptance of my wretched dermatitis-blighted body.

Energy surged through every cell in my body. All my senses came alive. I touched my lips to hers, and she began kissing me back. Our tongues danced together, and our bodies pressed closer. My hands explored her waist and reached down to circle her butt. She pulled at me more passionately—and then, she pushed me away.

"No," she said. "We can't do this."

"Okay," I said, gasping for breath. "It's all right. I understand."

To my surprise and confusion, Jessica cuddled right back up to me, her body and face pressed against mine. I leaned in again and touched my lips to hers. Once again, she responded, her arms pulling me closer. Then, she pulled away again.

"I told you," she said. "We can't do this. So don't be starting up again."

My head was spinning. After a time-out, our bodies drifted together again, and I went in for a kiss. She bolted upright in the bed and in a firm voice said to me, "Look, if you touch me one more time, you're going to have to leave the room. I'm serious. You can sleep here but you can't kiss me." I heard a soft chuckle from Kenya's bed across the room.

I didn't get much sleep that night. The next morning Jessica woke me at dawn so I could sneak out of her dorm undetected.

I didn't see her again until chemistry lab a few days later. With a goofy grin planted on my face, I sauntered over to chat with Jessica and Veronica. Jessica quickly shut me down. "Can we just keep it about chemistry lab? We have a lot of work to do today, and I don't have time for your jokes."

THAT WEEKEND, I once again needed a hideout. This time I approached Jessica more discreetly, out of Veronica's sight and hearing, to ask if there was space for me in her room that night. She said there was.

I made sure to take a shower that evening before I snuck upstairs to Jessica's dorm room and knocked quietly on her door. She answered immediately, dressed only in a T-shirt. When she beckoned me inside, I noticed that Kenya was not in the other twin bed.

"Should I sleep over there?" I asked.

"No," she replied. "You can sleep with me."

The moment I got into her bed, any lingering doubts were erased. She cuddled up to me as she had before. But this time, it was her lips that came searching for mine. My mouth hungrily devoured her kisses. Our hands roamed each other's bodies. When she reached below my waist, I knew it was on. She tore at my clothes, and I pulled at her body.

I'd had sex before. But I'd never truly made love. Not like this. Jessica was clearly much more experienced than me. I thought about Lisa and the promise ring I'd given her—sweet Lisa, innocent Lisa—and I told myself that this encounter would make me a better lover and husband to her someday.

As I grasped Jessica around the waist and pulled her close, she whispered in my ear. "Slow down. Let me show you."

I did let her, and she did show me.

Later, as we lay in a tangle of sweaty limbs, I felt reborn. So *this* is sex, I thought, as if I'd accidentally discovered the hidden Shangri-La of sensual abandon. Most amazing of all was what happened next. We cuddled. We chatted. We told secrets and whispered sugar in each other's ears. We made love two more times before dawn.

WE WERE STILL ASLEEP, all twisted up in the sheets, when a loud knocking woke us up.

"Open the door!" shouted a voice we both recognized as Denard's. We froze in place.

The night before, between lovemaking rounds, Jessica had told me her arguments with Denard often led to physical brawls. "We fight like cats and dogs," she'd told me. "I'm gonna dump him."

Denard pounded more forcefully on the door, which luckily was locked.

"I'm not opening," Jessica shouted from the bed.

Meanwhile, I was looking around for a place to hide my bare ass. There was no closet in the room, just a standing wardrobe that I instinctively crouched behind. I looked out the window, trying to calculate the velocity of a Plummer-sized mass falling three stories to the ground. I didn't enjoy cowering naked and afraid in front of the first woman I'd truly made love to. But I wasn't brave enough to confront Denard, or to subject my body to Newton's second law of motion.

"Plummer!" Denard shouted. "You in there?"

Jessica made the zipped-lip motion—as if I needed her coaching to stay quiet. Denard kept pounding on the door, and Jessica kept shouting back that he better get his ass out of the hallway before the campus police showed up and hauled him off to jail. I prayed she'd stop being so sassy. That door was just a piece of framed-in plywood, nothing a pissed-off jock like Denard couldn't kick in. I'd seen plenty of fights between Mama and her boyfriends that started with shouting matches through locked doors and ended in mayhem, tears, and splintered doorframes.

Denard finally decided to stop pounding on the door and save his hands for the upcoming basketball season. Before he left, he shouted through the door, "I know where you live, Plummer. You better be watching over your shoulder for me!" I climbed into my clothes and waited a safe interval before venturing out into the hallway.

AFTER I "WENT OVER" the next week—removing the painted wooden scroll from around my neck and becoming a full-fledged Nupe, as Kappa brothers were called—Jessica and I finally went public. Her friends were mostly okay with me, now that I'd made Kappa Alpha Psi. Some of her sorority sisters warned Jessica that I wasn't right for her, that I dealt weed and hung out with thugs that smoked. But we were a KAPPA-AKA couple and a hot item. I had a line of brothers at my back and the homecoming queen on my arm. College life was looking up.

41

EVEN IF I DID get carried away with my passionate relationship with Jessica, Lisa was still the woman I loved and planned to marry. Like most of the guys I knew at Tougaloo, my hometown girl and on-campus flings existed in parallel but separate universes. And yes, it was convenient for me, I guess, to be able to put my feelings for Lisa on ice while I was away at college, and to keep Lisa and Jessica in the dark about each other. But I wasn't confused about my priorities. Lisa was my true love, and my future happiness. Jessica was my red-hot, high-stepping girlfriend from Natchez.

I had experience balancing double lives before, code-switching between juke joints and science fairs, between thug and computer nerd, between country boy and frat brother. I figured I could manage some crosstown traffic in my love life.

I'D JUST ARRIVED BACK HOME for Thanksgiving weekend when Lisa dropped the news on me: she was pregnant. She had been for a while, going back to the beginning of summer vacation.

My first thought was: We only did it a couple of times and I pulled out! My second thought was: Please don't turn both our lives upside down by keeping this baby. But I didn't just come out and say those things. What I said was, "What are we going to do, Sweet?"

Lisa just looked away, and we sat there on the Morgans' front porch, thinking our thoughts and saying nothing. After a week of conversations that went nowhere, it became clear that Lisa wanted to keep the baby. That's when I knew I'd have to drop out of college so we could get married. If I had any confusion on that point, Lisa's mama, Miss Banny, made her expectations clear.

"Ain't none of my kids had kids before they got married," she told me. "My Dwane married Bridgette. You gonna marry Lisa?" I was driving Miss Banny to the drugstore at that moment, and I was re-lieved not to have to meet her glaring eyes just then.

"Yeah," I said. "I'm gonna marry her." And I meant it. I loved Lisa and I wanted to do the right thing by her. I'd given her a promise ring when I was still in the navy, and that was my pledge to marry her. I just wished it had worked out so I could finish school first. There was simply no way I could marry Lisa and stay in school. I'd have to drop out, move back to Heidelberg, and find some sort of work.

Meanwhile, I was hearing the opposite point of view, expressed emphatically by Mama and Bridgette. Both had been good students before they dropped out of high school when they got pregnant at sixteen. Their babies stole their education from them. And a lot more. They never gave me advice about my life, but they always thought I'd do something special, and they let me know loud and clear that Lisa wasn't the one I should sacrifice my future for.

Mama said, "She ain't the one, son. Do *not* marry her."

Bridgette said, "Brother, you just started school. You're going somewhere in life. Lisa's not the girl for you."

They both said to me: We live in this community. We can help take care of your baby and Lisa. Send her and the baby money, but stay in school.

Even Lisa told me, *No, don't drop out of school.* But she wanted the baby and she wanted a proper wedding ring.

This was the first time Lisa and I had to deal with something complicated together. Now that there were three of us involved, or at least two and a half, all our easy talk and planning seemed to evaporate. I assumed I'd marry Lisa, but I didn't want to be pressured by Miss Banny into doing it right away, before the baby was born.

After I didn't promise to marry Lisa before the baby arrived, the iron door shut against me. Lisa's mama and sister were constantly throwing shade and poisoning Lisa's mind against me. When I returned to Heidelberg for Christmas vacation, they treated me like a big-city college boy who questioned Jesus and knocked up their little girl.

I thought I could still make things right with Lisa. After all, we loved each other and wanted to be together. Then one afternoon she picked a big fight over something tiny. She'd asked me to show up at her house at five o'clock sharp, and I came at 5:10. That was it. She just shut me down. I came around every day and sat there next to Lisa, and she wouldn't say a single word to me. That went on for two whole weeks.

Back at Tougaloo in the New Year, I finally got up the nerve to tell Jessica I was going to be a dad. She was another strong woman I wasn't looking to lock horns with. I braced myself for a bruising tirade—or worse, another emotional shutdown. But Jessica surprised me.

She asked me two questions: "Do you still love Lisa?" and "Are you still intimate with her?" I leveled with Jessica. I told her I still loved Lisa and hoped to work things out with her—but she had totally shut me out and wouldn't even speak to me. Jessica joked that as long as I wasn't intimate with Lisa, our sex was too good to give up.

The next night I showed up at Jessica's dorm room to find her in tears. A group of her busybody sorority sisters told her to leave me because I had a secret baby back in Heidelberg. Jessica told them off, saying I'd already shared that information with her, and she was standing by me. But once she was alone in her room, she'd gone to pieces.

Jessica did stand by me. I was starting to feel like we had something going on that was more than just good sex.

MY SON MARTEL was born in February at the free hospital in Laurel where poor Black folks went to have their babies. It was a long, hard delivery. They had to use forceps to get him out.

I was the first person to get to the hospital. I couldn't get inside the ward to see Lisa and the baby—they'd wheeled her into a sterile surgical unit for the delivery so we yelled to each other through the door. "Go get me a pizza!" she shouted.

"You got it!" I shouted back, relieved that she'd finally spoken to me—even if it was just a food order. Now that Martel was actually born, the idea of Lisa and me being parents and raising up our son together made me happy.

By the time I got back with the pizza, other family had arrived and were crowded into Lisa's room. The baby was still in the nursery. Bridgette said, "Let's give these lovebirds some time by theyselves," and shooed everybody but me out of the room.

Once we were alone, I reached down and took Lisa's hand to kiss it. She immediately pulled it back. "You weren't there for me when I needed you," she said, without even looking up at me. "You don't need to be here now."

At that moment, my heart fell off a cliff and hit the ground hard.

42

WHEN I WALKED out of Lisa's hospital room, JG was waiting for me in the hallway. He flashed a big smile and tried to dap me up. "A son! Plummer, my man, we gotta celebrate."

I had no smile for JG. I was hurt and confused. How could Lisa still be fighting with me? Our son had arrived. The right thing to do was to drop any fussing between us and come together. But that's not what was happening. She'd pushed me away and I was in agony. All I knew is that I wanted the pain to stop.

I'd begun a ritual at Tougaloo of buying a fifth of liquor after finals to drown my sorrows. I wasn't much of a drinker, but failing to make the grade at Tougaloo filled me with shame, and I needed to dull the pain of a crappy report card. But the pain I felt that night when Lisa cast me out of her life and my son's life was deeper and stronger than anything alcohol could numb.

"Let's go to my daddy's," I said to JG.

I stayed quiet during the two-hour drive to New Orleans, while JG tried to pick me up. "Plummer, man, I know you say she been tripping. But now that y'all's got a son together, ain't no way she keeping that shit up. Watch. She gon' come around."

I just stared out beyond the headlights into the darkness.

"Babies got a way of changing shit," JG insisted. "Hell, even racists open up they hearts when they get a half-Black grandbaby. Hah!" Again, I didn't laugh. "Damn, Plummer! I ain't never seen you like this. What the fuck she say to you?"

"She dumped me, man."

"Hell naw! Ain't no way? You ain't reading her right."

"No other way to read it. She said, 'You wasn't there when I needed you, so I don't need you now.'"

"Damn! I wish we had some weed, Plummer. I know you could use a smoke right now. You want me to stop and get us a couple of forties for the ride?"

"Naw, man. Just keep driving. I know what I need."

IT WAS CLOSE TO MIDNIGHT by the time we pulled up to Daddy's house. He answered the door in a white sleeveless T-shirt and a wide-brimmed white hat.

"Hey, Na!" he called out, stepping forward to give me a hug, "Come on in. How you doin', JG?" he asked, dapping JG up.

"Hey, Mr. Plummer. You a grandpa! James's son was born today!"

"Dat right?" he looked to me. "Dat's won'erful! This your first, right?"

"Yes, sir," I said grimly.

"Come on in and have a seat. Wanna beer? Y'all wanna smoke a jernt?"

"Yeah," I answered. "But what I really want to smoke is some rock."

Daddy's face went from cheerful to serious. "Really? You smokin' rock now?"

"Shit's too heavy for me right now. I just need to feel good, or not feel anything at all."

Daddy sat us down at the table and handed us each a Budweiser. "Tell me what's goin' on."

I gave him the update. He nodded. "I understand," Daddy said. "In boxing, they say the punch that knock you out is the one you don't see comin'."

Then he turned to JG. "What y'all lookin' to get?"

"How much is two eight-balls?" JG asked.

"I can get you dat for $250 to $300," Daddy answered.

"Let's do it," JG said.

"A'right. Lemme make a call."

TWENTY MINUTES LATER the doorbell rang. JG and I waited at the dining-room table while Daddy transacted business in the living room. Then he came back and placed two plastic bags bulging with white chunks and powder on the table.

"Damn! Them some fat eight-balls!" JG said.

"Taste it," Daddy said.

JG dipped in and snorted a sample. "Wow! That there's the p-funk!"

Daddy looked to me. "You wanna taste?"

"No, sir." I needed something stronger than blow. "I'll wait for the rock."

"I'll cook it right up then." Daddy put a pot of water on the stove and came back to the table with a bottle of rubbing alcohol and a cigar box. Inside the box was a glass pipe, a couple one-sided razors, a piece of broken wire hanger, cotton balls, a candle, and a lighter.

Daddy set to cooking up rocks on the kitchen stove. He put a mound of coke in a beaker, and added water and baking soda. Then he placed the beaker inside the pot of hot water on the stove and swirled the mixture around. Within minutes, the solution went from cloudy to clear, leaving a single solid white ball at the bottom. Daddy poured off the water and the rock landed with a clink on the bottom of the sink.

Daddy put the rock on a small mirror, then used a razor blade to slice off a little chunk. "You ready?" he asked me.

"Yeah," I replied.

Daddy pressed the chunk into the glass pipe and, using a torch made from a wire hanger with a cotton ball wrapped tightly around its tip, melted the rock into a wad of steel wool packed into the bowl of the glass pipe. Then he handed the pipe to me.

"Here ya go, James Plummer Jr. . . . take it real slow and steady."

I raised the pipe to my lips and Daddy touched the torch to the bowl. I sucked the thick stream of white vapor into my lungs, and he moved the flame away. "Hold it for a bit, then let it go." I did like he said.

The cocaine raced across my blood-brain barrier and washed over me with a warm wave of feel-good. I exhaled all my stress and all my pain. Every cell in my body woke up. My brain hovered up above my head. I suddenly became aware of my breathing. I inhaled deeply and exhaled with a sigh.

"Wow!" I said, repeating JG's exclamation.

Daddy slowly nodded as if to say, *Yeah, you got it*. Then he cut off a piece of rock about twice the size of mine, hit it with the torch, and inhaled the vapor into his lungs. Then he loaded up the pipe for JG.

For the next six hours we three sat at the dining-room table smoking rocks.

JG and I talked nonstop between hits, and all of it was nonsense. I felt great. Better than I'd ever felt before. Like I'd injected jet fuel and was cruising at 30,000 feet—as long as I kept hitting that pipe.

At some point, I realized I was clenching my teeth real hard and Daddy was making a weird jerking movement with his mouth. His eyes were almost all black pupils, with just the thinnest ring of blue-gray irises still present. Every ten minutes he'd get up to peer through the window blinds into the darkness before coming back to hit the pipe again.

Sometime around dawn, Daddy's young wife, Stephanie, came downstairs. She was nine months pregnant and looked it.

"James," she said, tugging on Daddy's tank top. "I'm goin' into labor. You gotta take me to the hospital."

43

TWO WEEKS LATER, Bridgette called me at school to say that something was wrong with Martel. "You need to get down here," she said.

My chest seized up in a way it never had before I was a dad.

"They say Martel's got cerebral palsy."

"What's that?" I asked.

"They say a part of his brain is dead."

When I went home to see him that weekend at Lisa's, Martel's eyes wouldn't stop moving. They kept rolling around in his head.

"Look, this is killing me," I said to Lisa. "Martel needs two parents to take care of him. If you give me the word right now, we'll get married and be a family together." Her only reply was stone-cold silence.

I wasn't ready to give up. I showed up every weekend and sat with Lisa and Martel at her house. Lisa still wouldn't talk to me, but at least she let me take Martel back to Bridgette and Dwane's trailer overnight on Saturdays.

A few months later I heard from Bridgette that Lisa had moved to Houston to live with relatives—and left Martel with her mother in Heidelberg.

ALL THAT SPRING, I traveled back to Heidelberg to visit Martel. Jessica came along with me. As a nursing student, she understood cerebral palsy better than I did. She said that it was usually caused by birth trauma, or a stroke while the baby was still in the womb. The hospital staff insisted they'd done nothing wrong. We were poor Black people at the free hospital. It was easier for us to question the existence of Jesus than to challenge a doctor for an explanation of what might have gone wrong during my son's birth.

Jessica was a champion. Together, we began campaigning to have Martel come live with us. But in our matriarchal community, the rule pretty much held that the mother owns the child. Going to a judge or social welfare office for help would mean bringing white folks into a family dispute—which wasn't done.

When Martel turned three months old, Lisa returned from Houston and finally broke her silence—just long enough to tell me she had a new boyfriend. A dark cloud of failure and despair enveloped me. I couldn't be with Lisa and Martel as a family. I couldn't be a real father to my son. I was ashamed and miserable. I responded the only way I knew how—by self-destructing.

That semester I had an on-campus job tutoring math. All of my earnings went to buying coke. Every weekend, JG and I would take our money, hit the streets of Jackson to score, and spend a couple of days bingeing till our stash was gone. It took me a whole day to recover from the beating our bingeing put on my body. Meanwhile, my normally upbeat personality turned super sad. My academic performance declined as well. Jessica saw the change in me. But since I kept the crack binges secret from her, she chalked it up to what I was going through with Martel. It was practically all I talked about.

As summer loomed, I had no idea where to go or what to do. I knew I wasn't going home after classes ended in May. I didn't want to be a burden to my family, and there was nothing but sad times and unemployment back home anyway.

The only job I could find near Tougaloo was working at Wendy's.

Young Black men in Mississippi didn't work the front of the store in those days, not even in fast-food joints. I worked the fryer, emptying twenty-pound bags of frozen fries into vats of spattering grease and flipping those weird square hamburgers, which were Wendy's signature patties: "At Wendy's we don't cut corners." They paid minimum wage, which was $3.35. Wendy's gave us one free meal a shift, so that's the one meal I ate each day.

After sending Lisa $35 a week from my mini-paycheck, I couldn't afford to pay for housing. I heard from JG, who had a summer job on campus, that they'd left the air-conditioning on at one of the women's dorms that summer because they were renovating parts of the building. I found a way to sneak inside, and located a room with a mattress to sleep on.

One day in August, I was repackaging hamburger meat when I cut my hand on the serrated edge of an industrial-sized roll of plastic wrap. My hand immediately got infected. With no money and no insurance, I didn't go to the hospital for three days. By that time my hand was puffed out like a balloon, my lymph nodes had swollen up all over my body, and I was in horrible pain. I eventually found my way to the emergency room of the charity hospital in Jackson. They drained the pus out of my hand and shot me full of antibiotics. The doctor told me he didn't know if they could save all my fingers.

After a few days, the swelling started to go down. But by then, they'd given my fryer job at Wendy's to some other fool. I spent the last weeks of the summer in the women's dorm nursing my hand and waiting for my junior year to begin.

44

Back in May, it had seemed like a good idea to room with JG in the fall. He was waiting for me in our dorm room the first day of the term, all pumped up about his plan for how we were going to get rich together. I was dead broke. But JG had more than a thousand dollars in summer pay burning a hole in his pocket, and in his brain.

"Plummer," he said, pacing in circles around our double. "I bet we can triple our money if we buy some coke, rock that shit up, and sell it. I've got enough bread to buy a coupla ounces. Your Daddy, Mr. Plummer, be getting that p-funk—so I *know* we gon' sell out quick."

"Yeah, but what if we *smoke* out quick? You think we can hold that much rock and not smoke it all?"

"I didn't say we ain't gon' smoke none. But we damned sho' ain't smoking whole ounces. Just think, if we triple our money a couple times, we can up our quantity to a kilo. We could be pushing Caddys by the end of the year."

I didn't care about Caddys, or anything else on four wheels. But for someone who'd never made more than minimum wage, tripling our money seemed like the most plausible way to finally make some bank. But I knew that selling rocks meant a totally different, and much more dangerous, customer base than weed. Weed buyers were chill. Rock buyers were fiends. I weighed my primal fear of rock fiends against my desperation to make some grown-up money to send to Lisa and Martel.

"Fuck it. I'm in," I replied.

JG laughed. "That's what I like about you, Plummer. If I say, 'Let's . . .' it don't matter what come next. You be down as hell."

Even while we were dapping each other up over our new business partnership, the smarter part of me knew we was jive-talking ourselves. The only reason JG and I had never smoked ounce quantities of crack before was because we'd never been holding ounces before. The rules of rocks were as immutable as the laws of physics. They determined what *must* happen. Rock rule number one: Once you started smoking, you can't stop till it's all gone.

Scoring was the easy part. Daddy went in on a half-pound buy with us for $2,500. He kicked in $1,250 for his quarter pound, and JG fronted his summer earnings for our quarter. Daddy took us along to make the pickup, and to be his backup. He gave JG one of his pistols to tuck into his waist, and he tucked another into the small of his back. When he checked out my gun, which used to be Mama's, he smiled and said, "I remember the day I gave Laney that little piece. Dat was back when we was living together in the Goose, 'fo' you was even born."

Daddy didn't have to tell us to keep our heat out of sight. We understood you don't pull a gun out to show it. That just escalates a situation. You only pull it out to shoot, and only if you have to.

JG and I stood next to the car while Daddy met the man across the street at a garage. We watched the whole deal go down. When the guy handed Daddy the brick of coke, it wasn't wrapped or nothing. It looked like a large chunk of Styrofoam.

When we got back to Daddy's place and sliced into it, it was all shiny like white gold you'd pan from a river. Just seeing coke that pure made my heart race and my stomach lurch. When I rubbed a piece between my fingers it was smooth like butter, not chalky like the cut shit I'd handled before. JG rubbed some on his gums to get a freeze, and whooped his approval.

Daddy rocked a bit up and we took a couple of hits. Pure nitro. Then he weighed out our four ounces, packaged it up, and told us where to hide it in the car in case we were stopped on the highway. We took a few small rocks and a glass straight shooter so we could smoke during the drive back to Tougaloo. That's the thing about rocks. As soon as you're holding 'em, you gotta smoke 'em.

On our way through Jackson, we stopped on the ghetto side of town for supplies: rubbing alcohol, cotton balls, straight razors, lighters, a glass pipe and screens, and a whole lot of baking soda. Back on campus, I dipped into the chemistry lab for some test tubes, beakers, and a digital scale—strictly on loan. By nightfall, we'd kitted out our dorm room for cooking. Minutes later, our first rock rolled out of a beaker and onto the table.

"Damn, Plummer!" JG said, "That shit look proper!" He cut the rock in half, loaded it into a glass pipe, and hit it. We had lift-off.

WE SMOKED STRAIGHT ON THROUGH till sunrise, when we finally paused to rest. We woke at noon, feeling like roadkill. After showering, shaving, and changing clothes, we felt like clean-shaven roadkill. Over lunch at the cafeteria we vowed to only smoke in the evenings after dinner. Just to be on the safe side, we also vowed to only smoke on Thursdays, Fridays, and Saturdays. Starting next weekend. 'Cause it was Sunday, and we'd just scored.

And just to be sure we didn't overplay our hand, we decided to lay off quarter ounces to a friend of JG's at wholesale rather than selling it ourselves. Less profit that way, for sure, but we'd reduce our risk of nasty encounters with hardcore rock fiends on the mean streets of Jackson. If his friend didn't have the money to buy from us, we'd front it to him on consignment.

That afternoon we fronted a quarter ounce to JG's pal, just like we planned. We had this. Then we went to dinner at the dining hall, JG with his friends and me with my Kappa brothers, just like normal. After dinner, we went straight back to our room and began smoking again.

We spent the next three days and nights smoking rocks in our room. We didn't plan to. It just turned out that way. By the third dawn, we could smell the coke coming out of our pores. We'd gotten paranoid as hell, pausing to listen at the door at four in the morning, then at 4:07 and 4:12. We lived like vampires, terrified of daylight. We'd make mad dashes across the hallway to the bathroom for cooking water, but only late at night when everyone was sleeping. We kept a rolled towel wedged into the bottom of the door to keep the smell from leaking out, and we blew the smoke out the window through a paper-towel tube stuffed with fabric-softener sheets. Every couple of hours we'd take a break for fear our hearts might burst. We'd lie on the floor for a few minutes to try to slow the runaway trains in our chests. Then we'd get back at it.

Being holed up in our dorm room for days at a time, we completely fell out of campus life. Weekends became the only time we *didn't* smoke—since weekends is when we had to connect with our girlfriends. Jessica had moved to Jackson for nursing school, where she lived in a dorm at the University of Mississippi Medical Center. She spent her days studying and most all of her evenings working at the medical center. It was relatively easy for me to hide my rock habit from her, since we only saw each other on weekends. We divided our time together evenly between passionate lovemaking and intense fighting.

My frat brothers were closer at hand. We'd see each other at meals in the cafeteria where we all sat at the Kappa table. I was able to show up for most of the dinners and tell them I was focusing on my studies at night and spending my weekends with Jessica.

Keeping up with academic work was impossible. After a night of smoking, I was too tired and wrung out for morning classes. I'd read my textbooks. Or at least I'd open them for a while before dinner. But my already crappy study habits were no match for the mighty glass pipe. JG and I both fell behind academically. It was the beginning of my junior, and JG's senior, year. When midterms came around, we paused to tally our losses.

The bottom line wasn't pretty. We'd barely broken even on our big drug deal. After we ran through our stash, we started buying back rocks at retail from the same dudes we'd fronted them to at wholesale. Meanwhile, we were both pulling D's and F's in all our courses. There was no way I was going to be able to pass a single class. It was better to drop out, I figured, and not have the F's on my record. That way I could at least come back to school later and still be eligible for financial aid.

To get an approved leave of absence, we needed to go through official channels. That meant getting a signature from each class's professor, and sign-off from the academic dean, Dr. Bettye Parker-Smith. To run that gauntlet, we had to spend the morning washing away our vampire look and the afternoon masquerading as humans.

JG met with the dean first. Dr. Parker-Smith tried to persuade him to stay in school and finish up, since he was a senior. Why drop out a semester and a half before graduating? she reasoned. JG held his ground.

When it was my turn, Dr. Parker-Smith perused my transcript like it was a rap sheet. She asked me about an incident sophomore year that appeared on my record—a fight where I'd hit the freshman class president, Roger Horton, forcefully in the head with my Kappa cane. When I explained that Roger Horton had preemptively sucker-punched my frat brother, she raised an eyebrow and made a note on my transcript. Apparently, I hadn't showered all the evil vampire out of me, because I heard myself schooling the dean, at steadily increasing volume, about the Kappa code of honor that required me to crack open the head of any muthafucka who raised his hand against a brother.

"I see," she said, setting aside my transcript and looking me in the eye. "I think you're doing the right thing to drop out. College is clearly not the right fit for you now. Perhaps after some time away, and some personal reflection, you can mature a bit and then continue your education—either at Tougaloo or at some other institution."

I was offended. The dean seemed oblivious to my inner brilliance. If there was anywhere I belonged, it was college! The dean's job, as I saw it, was to persuade me to stay in school and realize my limitless potential. Instead she signed my paperwork and pushed it across her desk toward me.

"Good luck, Mr. Plummer," she said. "Please turn in your student ID and your meal card to the registrar—and be out of your dorm by tomorrow afternoon."

45

JG AND I found a furnished room at the Grove Apartments, half-way between Tougaloo and Jackson. The Grove was so low-rent, we didn't even have to put down a deposit. JG heard that the Ramada Renaissance, a fancy hotel off I-55, was hiring banquet waiters, and we beelined it over there.

The head of housekeeping, Miss Jolly, explained to me that if I worked for housekeeping instead of banquet waiting, I'd be guaranteed steadier work at $4 an hour. Since my priority was reliable cash flow, I went for the steady money and signed up to be a houseman—which was basically a janitor. JG was hired on as a banquet waiter.

Miss Jolly and the assistant manager were white. The entire housekeeping crew was Black. Miss Jolly was the type of white lady that you'd imagine running the house staff of a fine Southern mansion. She addressed all the staff as "boys" and "girls," even though we ranged in age from twenty to fifty.

I was a slow learner, as usual, when it came to subordinating my ego to authority figures. The Black maids often fell into friendly debates with one another about various matters of fact. When they reached an impasse, they'd call Miss Jolly to resolve the argument. Once, I overheard her spouting the wrong answer, and I jumped in to correct her. One of the maids said, "James went to Tougaloo, so he must know right."

Miss Jolly just smiled. She knew she could always even the score.

It didn't take her long to show me I wasn't a college boy anymore. A few days later, the Tougaloo administration was hosting an alumni event in the hotel reception hall. Right before the reception began, Miss Jolly called me into her office.

"James, it's come to my attention that there's a wax buildup on the baseboards in the reception hall. I'm afraid it's going to require hand-scrubbing." She handed me a brush the size you'd use to clean your nails.

"I'll get right on it, Miss Jolly, as soon as the reception's over."

"No, I'd like it taken care of right away."

That's how she wanted me—on my hands and knees, dressed in my gray bell-bottom uniform with the purple piping down the legs and arms, while the Tougaloo faculty and alumni sipped their cocktails and ate their canapés. I made sure to keep my head down at floor level the whole evening for fear one of my professors might recognize me. But I learned an important lesson that night: nobody takes any notice of a Black boy on his hands and knees scrubbing the baseboard. He fades right into the woodwork.

AT NIGHT AFTER WORK, JG and I hit the hoods of Colonial Heights, Washington Addition, and Bailey Avenue. Ash Street was a particularly dangerous location, but always a dependable spot to score. On the bad nights—and increasingly, they were mostly bad nights— I felt like a species of vermin seeking something dirty to consume in the trashiest locations. We'd score and return to the Grove, smoke through our rocks, and then be back out on the street at two in the morning. And again at four. Some nights we kept at it till we had to show up at the hotel for work the following day.

The rock dealers we scored from were mostly users selling to support their own habits. They didn't carry their product in the street. Instead, they'd take your order and go grab their stash from a hiding place nearby, or else invite you to follow them to some shady-ass location to do the deal. If you gave them the money up front, after they promised to return with the product, you'd never see them or your money again. We had to be on guard at all times. Usually JG would do the deals and the talking while I stood watch behind him with my hand on the gun in my jacket pocket.

The more time we spent among the all-in hundred-percenters, the more we became like them—ruthless and ready to pull any dirty move just to score. The most scandalous time was the night JG stole from his girlfriend, Dorothy. He knew she'd just cashed her student-loan check, and when we went empty at three in the morning, he snuck into her crib and stole the cash from her dresser. I tried to talk him out of the idea. But after he ripped off her cash and turned it into rocks, I was right there next to him sucking on the pipe.

The most brutal and recurrent feeling in the life of a crack smoker is the moment the rocks run out. One night after we went empty, we were out cruising for a score in a car I'd borrowed from Jessica— her daddy's Chevy SUV plastered with "Jesus Saves" bumper stickers. Some scarecrow of a guy standing next to his car flagged us down.

"What y'all looking for?" he asked through our rolled-down window.

"You got a couple dubs?" Forty dollars was all we could afford after a long night of smoking, and we were badly tweaked out.

"I gotcha . . . hold on just a sec." He reached into his car and came out holding two fat rocks in his hand.

JG passed them to me. "Check these out, Plummer."

Just the sight of rocks was enough to set my brain on fire. I sniffed them, and they smelled weird. When I scratched one with my fingernail, it left an indentation instead of breaking off a chip. I scratched the other one. Sure enough, they were two balls of wax.

"Muthafucka, this wax!" I exclaimed. Here I thought I was going to get rock and this guy done fucked me up by feeding my hunger for nothing. Rage surged through my body. I stepped out of the car and threw the fake rocks right into his chest. I thrust my hand into my pocket and seized my gun, staring straight into the scarecrow's eyes. At that moment I knew that if the gun came out of my pocket, it was coming out spitting bullets. I was hoping the scarecrow would jump bad—simply tell me, *Fuck you!*—so I could kill him. I'd just thrown his rocks into his chest, so he'd certainly have to challenge me and call me a liar. But for some reason he didn't. His eyes showed fear. And the last ounce of humanity left in me kept my gun in my pocket. I needed him to help me kill him.

"Muthafucka, you trying to cheat me? Huh?"

"Naw, man, I gave you the wrong ones. By mistake. I wasn't gon' cheat you."

"Yes, you was, muthafucka."

JG knew what I was thinking. He jumped out of the car and stepped between us.

"JG, move out the muthafuckin' way, I'ma show this muthafucka."

"Naw, Plummer. Don't do it. Let's go. It's cool. It's cool."

"Fuck this muthafucka!" I said, lifting my gun hand from my pocket and pointing my finger into the scarecrow's face. "He tried to rip us off."

JG spoke slow and steady, while I kept my eyes locked on the scarecrow. "Plummer, this ain't you. C'mon, man. Let's go."

I finally lowered my hand. The scarecrow jumped into his car and peeled away.

"What the fuck's wrong with you, JG?" I yelled, pounding my hand against the hood of the car. "We shoulda popped that muthafucka! Goddamn!"

JG looked at me with something I'd never seen in his eyes before—fear.

"Plummer, you changing, man. You gon' shoot some strung-out dude over forty dollars? C'mon, now. That's some fiendish shit. That ain't you, man."

I knew he was right. Who was I becoming? A crackhead killer? That wasn't me. Or was it?

I felt like I couldn't survive much longer out on the streets of

Jackson, which were looking more and more like a zombie apocalypse. I was carrying a gun for protection all the time now, everywhere I went. It was a war zone.

I was tired. I'd been struggling on my own for so long, just trying to keep my head above water. I was adrift on a moonless ocean, and felt a wave of darkness about to engulf me.

46

PUSHING A VACUUM CLEANER at the Ramada Renaissance wasn't making me enough money to cover rent, rocks, and food. I started scavenging leftover food from the room-service carts people left outside in the hallway.

I'd gotten friendly with the bellhop, who made good tips carrying customers' luggage up to their rooms. He wore a nicely tailored bell-man's outfit that matched the hotel colors. He was part of the interior decor. And luckily, we wore the same size suits. On weekends, after he'd made $100 in tips and wanted to take his woman out, he'd say, "James, you take over. I'm outta here." I'd usually make twenty or thirty bucks before I finished up his shift at midnight. Just enough to fund my late-night entertainment.

When the bellhop quit to take a job at the new casino in Biloxi, I applied for his position. The manager of the hotel, Mr. Fortuné Jaubert, was the decision-maker on all hires that interacted with the guests. He was a courtly Southern white guy who'd taken a liking to me—probably because we had a lot in common. We were both from New Orleans, had both been in the navy, and had both played the tuba.

A few days after I submitted my application, Mr. Jaubert called me into his office and motioned for me to sit down in the chair across from his desk. "James," he said, "I want you to know that I've noticed how hard you work. I'd like to give you the bellhop position.

Unfortunately"—and here he looked down at a printed sheet on his desk—"Miss Jolly has given you a poor recommendation."

I didn't respond. What could I say?

"I trust you can appreciate the delicate balance required between the front and back of the house," said Mr. Jaubert. "Between Miss Jolly's domain and my own. My hands are tied. I hope you understand."

"Yes, sir, Mr. Jaubert," I said. "I think I do."

He knew I'd gone to Tougaloo. And he must have seen the pain in my face, because he leaned in closer to me and looked me straight in the eye. "I know how you must feel," he said. "I started off in a similar position and worked my way up. I can tell you this: without those humbling experiences I would never have learned how to become a gentleman."

I appreciated that he was trying to make me feel better, and that he was telling me he believed I too could someday become a gentleman. But as I left his office and walked back to housekeeping, I thought to myself, I can't move up from janitor to bellhop? This is it? Until then, I didn't have a firm plan about going back to college. But I knew I couldn't be a janitor the rest of my life. Mr. Jaubert might have worked his way up from housekeeping, but I was never going to. Not as long as the Miss Jollys were calling the shots.

It was time to switch things up.

IN EARLY NOVEMBER, I took Jessica out for her birthday. I was coming off a two-day bender with JG, so I was in ragged shape and the only date I could afford was a movie and a fast-food dinner. I didn't even have a present for her.

We went to see *Pumpkinhead,* a creature feature horror film about a demonic monster made from a corpse who goes on a killing spree. The hero—the good guy who's trying to kill the monster—finally realizes that *he* is Pumpkinhead, and that the only way to kill the monster is to kill himself.

As we sat on a bench outside the theater, eating our McDonald's Big Meals out of our bags, I tried to shake the Pumpkinhead story line loose from my head. It was much too close to home. Did I have to kill myself, I wondered, to escape the monstrous nightmare my life had become?

I was going too hard, too big, too much of the time. I was either gonna get killed or I was gonna OD. I didn't want to become a lifer, someone who was out there on the street all the time. But more and more I saw myself reflected in the eyes of the hundred-percenters I was scoring from at all hours of the night.

If I wanted to survive and stay out of prison, I needed to stay off of the streets, especially in the killing hours from midnight to six A.M., when all the crazy shit went down. I wanted to have an economic future, which meant reenrolling in college and doing well. And I needed to heal my spirit. I felt demon-possessed. Crack cocaine called out to me in my dreams and in my waking hours. I knew I couldn't answer that call and keep on living.

Meanwhile, Jessica was in danger of flunking out of nursing school. The last semester at Mississippi Medical Center, her grades had been borderline. She may have been a top-twenty student at Natchez High School and done okay at Tougaloo, but now that she was in school outside the Black community, with higher academic expectations, she realized her educational background was sorely lacking. I once saw an essay she'd written, and I was shocked. She wrote at what seemed to me a sixth-grade level.

Jessica and I sat in silence on the bench, fingering our soggy French fries and contemplating our equally unappetizing futures.

I turned to her and said, "Hey, you wanna get married?"

"What?"

"You wanna get married?"

"Are you serious?"

"I'm serious."

I knew I wasn't obvious marriage material—a college dropout working as a janitor with no prospects for advancement, and a seriously desperate drug life she knew nothing about. I couldn't bring myself to tell her about my dark life with JG. It was too shameful to confess.

Jessica said she wasn't sure if she was actually in love with me. I was honest with her and said I wasn't sure I was in love with her either. When we started out it was mostly lust that had powered our relationship. But we'd gone through a roller-coaster year together, including the whole Martel nightmare. We'd established another layer of love and respect for each other that we didn't have before. I told her that I had screwed up and I wanted to make things right. I told her I wanted to go back to school, marry her, and live a stable life together.

She told me she could see my heart, and she could tell I was trying to be good. "You're trying to do the right thing by Martel," she said, reaching out and squeezing my hand. "That's the kind of guy I want to be with. The kind of man I want to be married to."

WE GOT MARRIED two weeks later at the courthouse in Jackson. Just the two of us. Jessica was a go-to-church-every-Sunday girl, but she didn't want a church wedding. Not the way we both were just then.

I'd only met her parents once, and we'd had instant bad chemistry. Her daddy was seriously religious, and he basically thought I was Satan. And after I dropped out of school, her sorority sisters didn't have a high opinion of me either.

So our wedding wasn't a happy, joy-joy moment where we said, "Hey everybody, let's all be in on this. Let's all celebrate and spend money about this." That wasn't an option. We were both down-and-outers who didn't feel like we deserved to celebrate. We just wanted to survive, or else go down together.

47

Jessica found us a one-bedroom apartment in a safe business district of Jackson for $200 a month. We quickly fell into a routine. She attended nursing school at the medical center, where she made $5 an hour wheeling patients around the hospital. I took classes at Tougaloo during the day and worked my janitor job at the Ramada three days a week from three P.M. to eleven P.M and one weekend shift. I spent the rest of my time studying to make up lost ground in school.

I was determined to succeed at Tougaloo this time around. Since it was still hard for me to learn inside a lecture hall, I committed to working as hard as I could outside of class. I knew I'd never succeed in science without being strong in math. So I found an equally committed study partner in my calculus class, and together we stood at a blackboard and worked every example problem and every end-of-chapter problem in the book. I got an A in calculus that term, and more important, I learned the value of not just working through problems but *talking* through them with a partner—which would continue to be my best strategy for learning difficult material and solving problems in any subject.

As soon as I reenrolled at Tougaloo, I declared physics as my major—with a double major in math, just to make sure I took plenty of math classes before I graduated. Physics was my first love, intellectually, going back to my World Book beginnings. Even though I had been bingeing on crack for days at a time the previous term, I still scored 100 percent on every physics exam. My physics professor, Dr. Dave Teal, joked that he could use my exams as the solution key rather than taking the time to make one himself.

Dr. Teal was the only physics professor at Tougaloo, and there was only one other physics major in my grade. There hadn't even been a physics department at Tougaloo before Dr. Teal arrived in 1965, straight from Caltech and Harvard, where he earned his PhD in physics. The courts had just ordered Mississippi colleges to desegregate, and working through his church in Jackson, Dr. Teal got deeply involved in the civil rights struggle. Early on, some students questioned his motives, wondering if he was some sort of missionary who'd come to uplift the poor Southern Black kids for a term before going home to write a book about it. But like a lot of the other white science professors at Tougaloo, Dr. Teal came south for a few years and ended up spending his whole career there.

He spent a lot of his time teaching intro physics courses. But when motivated or gifted students came along, Dr. Teal made sure to teach them what he knew. Tougaloo didn't offer a course in quantum mechanics, which I wanted to learn, and I had too many scheduling conflicts to take a course being offered nearby at Millsaps College. So Dr. Teal pulled together a curriculum just for me, and tutored me through the material one-on-one.

Professor Teal was committed to keeping his students in school—especially students like me who were also holding down night jobs—helping them find fellowship funding and guiding them toward graduate programs. He also wasn't above tracking me down and dragging me to class if I didn't show up as scheduled.

One day early in winter term, soon after I reenrolled, he came looking for me at the dining hall during lunchtime.

"James, you're going to be at the library on time this afternoon, right?"

"F'sure, Dr. Teal," I said, trying to remember what I was supposed to be showing up for. "What time was that again?"

"The folks from MIT will be starting their presentation at three o'clock sharp."

When I got to the library at five after three, I found Dr. Teal sitting at one end of a conference table with the other five Tougaloo physics majors. Three people I didn't recognize—a young woman and two young men—sat at the other end.

"Now that James is here," Dr. Teal said, giving me a stern look, "we can begin. We have three graduate students in physics from MIT visiting today. They'd like to talk to you about a wonderful professional opportunity. So please give them your full attention."

I could tell right off that these Northerners were a different breed of science nerd. They were Black and urban, but not street, and certainly not ghetto fabulous in their dress. They looked more like our white science professors than students, and they sat up straight at the table. The woman wore a blazer and a pair of oversized glasses that were one part style and two parts nerd. I noticed she had long, elegant hands with rings on every finger.

"Thank you all for coming," she said, scanning the table to make eye contact with each of us. "My name is Cynthia McIntyre, and these are my colleagues Fuad Muhammad and Claude Poux." She spoke in a low-pitched, sultry voice that I guessed was how women science nerds up north talked. "When I started studying physics at MIT I had no idea there were other Black graduate students. In MIT's entire history, only one Black woman has earned a PhD in physics. I'll be the second. And let me tell you," she said, looking at the two women undergraduates sitting across from her, "there aren't many of us anywhere in the country.

"Then I met Fuad and realized that I was not alone. He introduced me to Claude. We put our heads together to figure out how to increase the number of Black students in physics graduate programs. We approached MIT's deans of science and engineering to help us bring a cohort of undergraduates to MIT next month for a conference of Black physics students and Black physicists. We're inviting you to be among them."

She paused, looking around the table for someone to respond to her offer. After an uncomfortable silence, I spoke up. "So y'all want us to come to a conference at MIT? I don't even know where tha's at! And I sure ain't got no money to get there."

The Tougaloo women at the table cut me a look like I was embarrassing them.

"Perhaps I wasn't clear," said Cynthia. "We will pay all your expenses to come to MIT—the Massachusetts Institute of Technology—which is just outside Boston. We'll buy your airline tickets, provide you with hotel rooms, and cover all your meals."

You could hear the gasps around the table. No one had ever offered us free flights or hotel rooms anywhere, anytime. We were all in.

48

WHEN WE GOT OFF our flight at Logan Airport, I got my first taste of a Boston winter. In Mississippi, we called what passed for cold winter wind "the hawk." When the double glass doors swung open outside baggage claim, the wind felt like a ptero-dactyl taking a chain saw to my chest while he chewed off my entire face. I couldn't get into the shuttle bus fast enough.

At our hotel we were each assigned roommates from other colleges. Mine was a guy from Lincoln University, an HBCU in Penn-sylvania that I'd never heard of. There were students from all over the country, about half of them from small HBCUs. The other half were from highfalutin HBCUs like Morehouse and from big state schools like University of Michigan.

MIT must have been an impressive campus under all the snow and ice, but it reminded me of Superman's frosted-over Fortress of Solitude. Inside the conference center we were greeted by Cynthia McIntyre, all tall and regal in her blazer and bigger-than-life glasses. She explained that the conference was designed to develop a stu-dent support network within the Black physics community, to raise awareness among Black physics students of graduate-level and pro-fessional opportunities, and to expose future grad students to new developments in physics.

"I encourage you all to meet each other and have discussions among yourselves over lunch. Later, we'll go to the lecture hall and hear scientific talks from some prominent Black physicists."

The whole scene was disorienting, what with the cold and the snow and the Northern accents and strange food. I poked my spoon around in my New England clam chowder and scanned the tables full of bright-eyed future physicists. I overheard a student behind me bragging about his first-author credit on a paper he'd submitted for publication, and another asking whether *Nature* or *Science* was a higher-impact journal to publish in. I turned my attention back to my bowl and tried to count the clams in my chowder, but they all seemed to be chopped into tiny pieces.

AFTER LUNCH, they ushered us into an auditorium to listen to a roster of Black physics professors give talks on their research into things like string theory, nonlinear dynamics, and mathematical modeling. I was hoping the quantum physicist from Bell Labs would talk about relativity. But he presented math equations I'd never seen and couldn't follow.

This was my first introduction to real academic physics. But it barely felt real to me. I couldn't understand 99 percent of what they talked about. None of us Tougaloo students could. These scientists in their three-piece suits were Black, but they were Black men from another planet. I couldn't imagine any of them had ever sold weed, plucked a chicken, or played dice on a corner. The same was true of Cynthia McIntyre and her crew. There was no hood in any of them that I could detect. I wondered if having hood in me meant that I wasn't cut out to be a physicist.

THE PANEL DISCUSSIONS that followed were much more practical. Even though I was a physics major, I'd never thought about the nuts and bolts of applying to graduate school, not to mention finding a career as a physicist. I was glad to hear that physics PhD students not only didn't have to pay tuition, they actually got a small salary to conduct research and teach undergraduates. But first you had to get accepted, which meant having a good GPA in physics and math courses, high scores on the GRE general and subject tests, recommendation letters, and research experience.

Almost nothing I had done up to that point qualified me for physics graduate school. My grades were poor. I had no summer research experience. I wondered if it was even worth all the hard work I'd have to do to get into any grad school, much less a good one.

THE LAST EVENT of the conference was a speech by astronaut Fred Gregory. I'd never seen an astronaut in person before, let alone a Black astronaut. I wasn't going to miss this.

When I got to the lecture hall, it was standing room only. Gregory began by showing amazing photos of his flights on the space shuttle. But what really inspired me was his speech. Gregory told us how when he joined the astronaut corps in 1978, he was one of three Black astronauts. Ron McNair, Guion "Guy" Bluford, and he were the first three Black men in outer space. Bluford first flew in 1983, McNair followed in 1984, and in 1985 Fred Gregory piloted the launch of Spacelab 3.

"When the three of us were selected, we promised each other that we would not retire until three new Black astronauts joined up and flew to space. We added one when Charlie Bolden became an astronaut in 1981. But then we tragically lost Ron in the 1986 *Challenger* accident. He was a great man, and we miss him terribly. Ron earned his PhD in physics—like many of you will someday—and he did it right here at MIT.

"NASA just selected its first Black female astronaut. Her name is Mae Jemison. She went to Stanford undergrad and got her MD from Cornell. Mark my words: you'll be hearing a lot more about her in the future. So Charlie and Mae make two. We need one more Black astronaut—and then I can retire."

He paused to scan slowly across the room of young Black physics students.

"Don't shrink from this moment. History awaits you. Get your PhD. Apply to the astronaut program. And for God's sake, keep moving forward. We need you. America needs you. The Black community needs you."

Everyone was on their feet, and I was clapping and hollering louder than anyone.

During the flight home that night, after they dimmed the cabin lights, I leaned my forehead against the window and gazed out into the night sky. I tried to imagine what outer space must look like from the window of the shuttle, how dark the sky must be, how bright the stars. I pictured myself suited up and stepping out for a spacewalk, free-falling around the Earth so fast I'd never hit its surface. It seemed like a perfectly natural place for me to be. Why not shoot for the stars?

THE NEXT AFTERNOON, I was back on planet Earth, vacuuming hallway carpets at the Ramada Renaissance. One of the cleaning ladies told me that JG hadn't shown up for work the past week and asked if I knew what was up with him. I'd been steering clear of JG ever since I married Jessica and moved in with her. The rocks still called to me, particularly at night, and I knew I couldn't trust myself to resist them and JG in combination. He'd been calling me up every week or so to see if I wanted to connect. Just hearing his tweaked-out voice on the phone reminded me of why I needed to go straight home from work at night.

But the news that JG was a no-show at work had me worried. I decided to look in on him to make sure he was okay—or at least still eating. Maybe I could talk him into going home to Heidelberg and taking a break from the life for a while.

I dropped by the Grove Apartments, where he still had a furnished room. JG came to the door totally cranked up. He reported with great excitement that he'd just sold his bed for $50, and asked did I want to score a couple of dubs with him and party?

I thought to myself: Damn, he's too far gone now for me to bring him back. Selling his bed for rocks? That's some fearsome shit. I peeked inside at his formerly furnished room and saw there was no TV and, sure enough, no bed. Just a pile of sheets and ratty blankets on the floor.

I told JG I was late to a study session, and I beat it out of there.

49

F I WAS SERIOUS about applying to graduate school, I knew I had to get some research experience. At the conference they gave us a list of summer research programs at NASA, Department of Energy National Labs, and at various research centers and universities. I applied to every one of them—and every one of them turned me down. My overall GPA was only 2.56, and that wasn't going to cut it.

I was feeling pretty grim about my job prospects for the summer, and beyond. Then one morning during final exams I woke up to someone calling from Tougaloo's administration office. "University of Georgia says they need your Social Security number before they can issue your first check."

I had no idea what he was talking about. But since "check" was in his sentence, I climbed into my clothes and found my way to the administration office. It turned out I'd been accepted into University of Georgia's summer research program in chemistry that was funded by the National Science Foundation. The thing was, I'd never applied to the program. I tracked down my chemistry professor, Dr. Richard McGinnis—the one who'd arranged my job as TA in General Chemistry.

"You know anything about this?" I asked.

"So, they accepted you!" he exclaimed with a smile he usually reserved for students who could successfully diagram a complex molecule. "A professor I know over there called and asked me if I could recommend anyone. I told him you were my most promising chemistry student and a great TA. You got in, that's awesome!"

"Thanks," I said, "but I guess it doesn't matter. I ain't got a car or no money to get to Athens, Georgia."

A year earlier, while JG and I were on one of our late-night binges in Jackson, I ran out of gas and parked my car in a bad hood overnight. When I went back to retrieve it the next day, it had been stripped to the chassis. I'd been taking public transportation or borrowing Jessica's car ever since.

"Tell you what," said Professor McGinnis. "If I buy you a Greyhound ticket, will you pay me back?"

"Absolutely! Out of my first paycheck!"

I couldn't believe that Professor McGinnis would do that for me. I'd grown up surrounded by a deep distrust of white men, and here was a guy who had not only lined up a serious research gig for me, but he was fronting me transportation out of his own pocket. Like Dr. Teal and a lot of the science faculty, Dr. McGinnis had come down to Tougaloo as a Freedom Rider and then decided to stick around. Why would all these white guys from Caltech and Harvard want to spend their careers teaching Black kids in a small HBCU most folks had never heard of?

It gave me something to puzzle over during the daylong bus trip to Athens. I had no idea where I was going or what awaited me. I'd never done research in a lab before. All I knew was it had to be easier and more interesting than mowing lawns in the summer heat or working the fryer in some fast-food joint.

MY FIRST MORNING at University of Georgia, I was greeted by my summer mentor, a chemistry professor named Michael Duncan. He showed me around his lab, which was dominated by a giant laser and populated by an all-white group of young researchers from across the country. He explained that his team was investigating ultracold small molecular clusters, and my summer project was to support their research. I didn't understand anything about ultracold molecular clusters, but he assured me I'd pick it up as the summer went along. "Professor McGinnis tells me you're a fast learner."

Then he opened an envelope and pulled out a ring of keys. "You're going to need to get into the building, so here is a key to the front door downstairs. Here's a key to the lab. And you're going to need to get into my office sometimes, so here is a key to my office."

I felt like I was in *The Twilight Zone*. A white man had just handed over the keys to his office, his lab, and the entire building! The lab was crammed with seriously expensive equipment. Lasers and all sorts of high-tech measuring equipment. No one had ever shown me that level of trust, whether I'd earned it or not. I'd always been treated—certainly by most white folks in authority—like a suspected criminal. Yet this guy trusted me. And we'd just met!

The next morning, I arrived for work at eight A.M. Nobody else even showed up at the lab till ten. Then they had a cup of coffee and did a little work—then it was time for lunch. There was no time clock to punch or even a log-in sheet. People just seemed to come and go at random hours. After a few days of this, I asked a graduate student, "What time am I supposed to show up? Because every time I get here in the morning, no one's here."

"Look, man," he explained, "it's not about when you're here. It's about getting the work done. Right? You're working alone, so work when you want to but get your work done and show up for the weekly group meeting. That's it."

At that moment a bright neon sign flashed in my loner's brain that read: RESEARCH IS THE FIELD FOR YOU!

This whole business of trust and autonomy was huge for me. Not only had I constantly been accused of being a thief throughout my youth, people treated me as though that's what they expected of me. I was expected to lie and steal. So I found myself lying and stealing. You expect me to be the intimidating thug? I'll be that. You think I'm scary and not to be trusted? Let me show you whose ass can rob you blind.

But not that summer. The moment I stepped into that lab, they assumed I was smart or else I wouldn't be there. After all, my chemistry professor at Tougaloo had vouched for me. I was *expected* to do my job. At the end of the day, I was going to be judged on what I actually contributed. That was an exhilarating and novel concept. It gave me the brain space to wonder, for the first time in my life, what kind of scientist I might want to be. Up till that moment, I'd been locked into day-to-day survival questions like How can I eat today? and Where can I find a place to sleep indoors tonight? But now I started to ask myself: What am I really good at, and what could I actually accomplish if I focused my mind on it?

DR. DUNCAN'S GROUP was conducting research on a level I'd never seen or imagined at Tougaloo. We were using lasers to liberate atoms from a target element. Then we cooled the atoms and molecules to near absolute zero with supersonic jet expansion until they clustered into nontypical structures. My job was to compute those nanoclusters using a molecular mechanics code called MM2. Then I compared the MM2 code's outputs to validated cluster measurements to determine the accuracy of the simulations. To find those measurements, I had to conduct deep dives into scientific journals—and I discovered I had a talent for lit searches! I could maintain focus for hour upon hour of tedious fine-tooth-combing through journals. And just like Dr. Duncan promised, within two weeks I actually understood this crazy shit!

I also learned that the OCD traits I'd been struggling with most of my life—the compulsive counting, the need to have things orderly and lined up in neat rows—were actually an asset in a research lab. I could hyperfocus on small, detailed tasks for hours. And for reasons I didn't fully understand, I could hold complex series of computations in my head, often seeing the end of an extended calculation as soon as I began.

I worked my ass off all summer to win Dr. Duncan's respect, and the respect of his grad students and postdocs. At the end of the summer they all encouraged me, telling me that I had the "it" factor to be a successful researcher.

Nobody in that all-white lab looked like me. Nobody handling those expensive lasers had my kind of childhood or had fucked up in the ways I had. And yet, they accepted me, and I felt perfectly at home there. Outside on the streets of Athens people in passing cars might yell out "Nigger!" or throw stuff at my head. But inside Dr. Duncan's lab, no one was out to get me. I didn't feel belittled or looked down on. I felt safe, and valued.

RIDING THE BUS BACK HOME at the end of August was like returning to Earth from a friendly alien abduction. I'd been beamed up to a faraway realm of futuristic science and technology—a world where I finally felt I belonged, where I was no longer judged by the social signifiers of color and class. Now that I'd gotten a taste of it, I was hungry for more—not just the space-age research and the camaraderie of a lab group. I wanted more of these new and unfamiliar feelings: the effortless immersion in a task I was good at; the ambition to stretch myself beyond what I knew I could do; and best of all, working hard to become the first person to find the answer to a scientific question.

Then the bus pulled into the Greyhound terminal, and I was back in Jackson with my shitty 2.56 GPA.

50

JESSICA AND I had a sweet reunion in Jackson. She'd been in summer school and was doing a lot better in her classes too. We both felt more hopeful about our futures than we had in years.

For the first time in a while, I was excited to start up classes again that fall. I was determined to get my grades up, and I finally had a strong faculty support team in place. In addition to Dr. Teal and Dr. McGinnis, I found another mentor in Gerald Bruno, who taught Tougaloo's toughest chemistry course, Physical Chemistry.

Dr. Bruno recruited me to work with him on a research project that involved programming integral equations to model electron-spin resonance. He encouraged me to learn to code in FORTRAN, which was a much more powerful language than BASIC. After we completed the project, he invited me to present our findings with him at a conference at Northern Arizona University.

During the two-day drive out to Flagstaff, we had plenty of time to talk. Dr. Bruno told me I had a special talent for problem-solving and research. I let him know that I was interested in applying to graduate school, but worried that I wouldn't get in anywhere because of my poor academic performance the first year and a half of school.

"Listen," he said, "you have what it takes. Your other professors all speak highly of your intellect and initiative. And based on what I've seen so far, I agree. It's not too late for you to turn things around.

If you put together three strong semesters, you can still get into a top program. You just have to continue to do well in your research, make some presentations like we're doing together this week, and maybe even get your name on a publication."

"Did you say get into a *top* program?" I asked.

"Sure. What really matters is your performance in upper-division courses. It's much better to get C's in your first two years and A's in your last two years. It's all about showing improvement and motivation. Research is what grad schools really care about, and you're a natural."

Hearing him say that was a huge boost. Dr. Bruno encouraged me to go to the upcoming National Conference of Black Physics Students and meet as many recruiters as I could from the schools I wanted to attend. "Just let them get to know you. Talk to them about your goals and ambitions."

Northern Arizona University had none of the stodginess of East Coast schools, and none of the rigid racial barriers of the South. A Kappa Alpha Psi brother at NAU invited me to watch a championship boxing match on pay TV. When I arrived at his house, I discovered it wasn't a fraternity gathering at all. There were forty or fifty people there, all having a good time. Black people, white people, Mexican Americans, Asians, and Navajo all sitting and partying together. For someone like me who'd always despised the racial divisions of the South, this calico-colored crowd was a thing of beauty.

51

THE FALL OF my senior year, I went to the National Conference of Black Physics Students in Hampton Roads, Virginia, hosted by the local HBCU, Hampton University. I was getting A's in all my upper-level courses by then, so I took Dr. Bruno's advice and focused on meeting recruiters from the top-tier graduate programs. There were recruiters and grad students from many of the best physics programs in the country. I interviewed with all of them. In the end, I applied to ten different programs, hoping I'd find a match somewhere.

I got accepted at four schools, including Stanford University's Graduate School of Physics. I'd heard that Stanford had graduated more Black PhDs than any other physics program. Their letter of acceptance included an invitation on stiff, cream-colored paper:

"We hope you can attend Stanford's Minority Admitted Students Weekend on April 19–21 being hosted by the graduate science departments."

At the bottom of the invitation, it said: "Please contact our travel office to arrange complimentary airline reservations." That's when I realized that things had flipped. Now that Stanford had accepted me, they wanted me to say yes to *them*.

There was nothing subtle about their courtship technique. I was met on campus by Homer Neal—a fifth-year Black graduate student in the physics department. Homer soon let me know that I was the only Black first-year physics applicant they'd accepted. He was a typical science nerd and almost a parody of a whitewashed Black guy. If you closed your eyes, there was no way you'd know he was a brother. In the first hour after we met, he'd confessed to me that he was still a virgin—and he was a fifth-year grad student! What Black dude would ever admit to that?

He showed me around the campus, which looked more like a country club than any university I'd ever seen. They even had an eighteen-hole golf course. The campus was filled with exotic trees I didn't recognize, with all sorts of squirrels running around their branches. I wondered why folks weren't hunting and eating all those squirrels, but figured if I asked Homer he'd probably tag me as a backwoods hick.

What really shocked me were all the coeds lying around sunbathing on towels like they were at the beach. It was a sunny spring day and every patch of lawn was wall-to-wall with white and Asian chicks in string bikini tops and tiny cutoff jeans that showed off their panties. To be honest, I was embarrassed to look at them. In the South, Black girls would never show so much skin in public. It just wasn't done.

Homer was keenly interested in particle physics, which was his obsession. The highlight of his campus tour was the Stanford Linear Accelerator Center. SLAC, as Homer called it, ran underground for two miles. It was the longest linear accelerator in the world, sling-shotting electrons at 99.99 percent of the speed of light. It wasn't much to look at from the outside, but Homer told me that Stanford researchers had won four Nobel Prizes for experiments conducted inside SLAC.

Homer and I sat out on the lawn overlooking the accelerator and debated a question only two physics nerds would give a shit about: Can protons decay? Recent cosmology models required protons to decay, but Homer insisted that they didn't. It was hard to argue with a fifth-year Stanford graduate student in particle physics, but I had an ace in the hole to play.

"When I attended the National Society of Black Physicists meeting this year," I told him, "a famous particle physicist gave a lecture where he said that protons *should* decay."

"That particle physicist would have been my father," said Homer. "And he would not have said that."

"Your father's a particle physicist?" I asked, trying not to sound overly impressed.

"Yeah. Homer Neal, Sr. He's the chair of the physics department at University of Michigan and a member of the DZero Experiment at Fermilab, where they're trying to observe the top quark."

I didn't have a comeback for that. I tried changing the topic. "So," I said, "you've got the same name as your daddy? Me too! I'm James Edward Plummer Jr."

"That's cool," said Homer, a model of politeness. "What does James Plummer Sr. do?"

"Ah, he works in an aluminum factory," I said. "Or he did until he retired." Now that I'd been admitted into the Stanford country club, I wasn't about to share war stories from my life on the street, or my daddy's.

Our next stop was the admitted science students "welcome" barbecue. It wasn't a real barbecue—certainly not by Mississippi standards. No red hots, and not even any ribs. Just a bunch of fancy "minority students" eating hamburgers and hot dogs with their pinky fingers practically turned up in the air. It didn't take me long to figure out that I was one of the few "deprived-Negro-from-the-hood" types—and the only one in the graduate physics department.

The next day, I was scheduled to meet individually with each of the professors on the physics faculty. When I asked Homer to brief me on them, they sounded mostly like the usual suspects—white guys from Ivy League schools, or else from universities in Europe, who all seemed to be gunning for the Nobel Prize.

The only professor of color in the department was a world-renowned solar physicist named Art Walker.

ART WALKER HAD been appointed to Stanford's physics department in the early '80s. That was after he became famous for designing x-ray telescopes that he launched into space to study the sun and the matter in space between the stars called the interstellar medium. "The spectra Dr. Walker's telescopes captured of the sun's corona revolutionized our understanding of the star at the center of our solar system." That's what it said in the physics department brochure that came with my acceptance package.

Walker was an only child of Black and Barbadian parents. His daddy was a lawyer, and his mother was a social worker. He grew up in Harlem, and his mama made sure he got a good education, including at New York City's best science high school.

After getting a physics BA at Case Western, and a master's and PhD in astrophysics at University of Illinois, Walker went to work at the Air Force Weapons Laboratory, where he designed rockets to launch satellites and cameras into space. Before coming to Stanford, he worked for a decade at the Aerospace Corporation studying the sun and upper atmosphere of the Earth. His most recent big-deal project was designing and launching multilayer mirrors into space that imaged the sun's extreme ultraviolet light and soft x-rays. He was a superstar in the world of solar physics.

Before visiting Walker, I made the speed-dating rounds of the other physics faculty. They were all happy to talk about their research without interruption. I listened as closely as I could, nodding as if I understood what they were telling me. They didn't ask me any questions. We were both grateful to escape as soon as our fifteen minutes were up.

When I got to Professor Walker's office, I could barely see him behind all the piles of paper stacked on every horizontal surface, including the floor. I stood in the doorway, staring at the top of his head, which was all I could see of him behind his barricade of science journals. He was hard at work with some kind of protractor, measuring a large blueprint spread over a wooden drafting table.

I knocked on the doorframe to get his attention.

"Yes?" he said without looking up at me. After continuing to focus intently on his measurements for another twenty seconds, he finally lifted his head and saw me. The furrow in his brow relaxed and he smiled.

"Hello! Come on in. Just, uh . . . move a stack somewhere and have a seat."

I moved a four-foot pile of papers from a chair to the floor, careful not to topple it into a heap.

"Welcome to Stanford," he said. "Tell me about yourself."

I narrated my not-so-impressive academic bio, skipping over most of my personal background and emphasizing that I was interested in space research. I told him about a summer research project I had lined up with a group at Berkeley that was constructing an experimental dark-matter detector.

"Very good!" Walker responded enthusiastically. "I run a program of experimental astrophysics using sounding rockets to image the sun's corona. This spring we'll fly an array of fourteen telescopes to study the sun's atmosphere. We design, calibrate, and launch our experimental payloads entirely on our own, so everyone gets hands-on experience."

I'd been coached by Cynthia McIntyre to ask two questions of any professor I might be doing research with: "How important is it to you that your grad students are gainfully employed after graduation?" and "What are your former students doing now?" Earlier in the day when I'd asked the other Stanford professors those questions, they'd looked at me side-eyed, as if I was getting way out ahead of myself. But Professor Walker wasn't like that. He answered me straight up.

"Making sure my students get good positions is very important to me," he said. "My first PhD candidate was Sally Ride," he said. "She's done just fine." He didn't have to tell me she'd been the first woman astronaut in space. He went on to tell me about several Black students he'd mentored who were working in research and industry.

Space research. Rocket-launched payloads. Telescopes. I was sold.

"I want to come work with your group, Professor Walker!"

He just chuckled. "We'll see."

I DIDN'T MAKE a final decision about which graduate school to attend until I found out how much funding they could offer me. When Jessica handed me Stanford's financial-aid letter that arrived in the mail, I held the envelope in my hand, afraid to open it. I wondered if I could ever fit in there with its whole country-club vibe. Art Walker was great, but if I went to Stanford, would I end up being the all-time champion code-switcher? And would Jessica ever feel at home there?

On the back of the envelope, the school's motto was printed in German: *Die Luft der Freiheit weht*, with the English translation below: "The wind of freedom blows." That seemed like a good omen. I opened the letter.

When we saw that Stanford was offering me a $15,000-a-year stipend—which was $4,000 higher than the other three schools—Jessica and I jumped up and down until the floorboards in our tiny State Street apartment began to shake.

52

BEFORE JESSICA AND I headed west to California, I wanted to go say goodbye to Daddy. He didn't make it to my Tougaloo graduation, and I hadn't seen him in a while. My older brother Byron gave me the address where he was living. The apartment was in a ghetto neighborhood in New Orleans East, not at all up to Daddy's standards. He'd always lived in a nice house in a decent, middle-class hood. Byron warned me not to leave my car on the street and out of sight for long. I parked in the lot next to the apartment house and made sure not to leave anything on the seats.

When I knocked at Daddy's apartment, someone shouted through the closed door, wanting to know who I was and what my business was. After he finally let me in, I was shocked to see it was Bobby Kelly from Piney Woods—our cousin who sold Daddy the B&M Club! He didn't seem to recognize or remember me, and to be honest, I barely recognized him all tweaked out like that. I could see someone crawling across the carpet behind him, examining tiny pieces of lint in hopes of finding some crumb of crack that might have fallen to the ground. It turned out to be his brother, Haskell Kelly. I had no idea what the Kelly brothers were doing in Daddy's apartment, but I'd heard that Piney Woods had gotten just as cracked out as Jackson and New Orleans East. Daddy was sitting on the sofa, his eyeball pressed against the end of a glass pipe in hopes of spying some residue worth smoking. It was that grim, hungry time of morning.

"James Junior . . ." Daddy mumbled, barely looking up from the pipe. "Good to see ya." His T-shirt was stained. He was unshaven and bleary-eyed.

"Good to see you, too Daddy," I lied. It near on broke my heart to see him looking like a hundred-percenter. Daddy and the Kelly brothers had always been big-timers. Hip, smooth dudes driving nice cars and wearing fine threads. When I was growing up, they set the standard for style and presence. They were men who always earned money. Now they looked like . . . like slaves, is what I thought. And it made me want to run straight out of there and back to my car.

"Wanna smoke a jernt?" Daddy asked, scanning the room for one. That's when I noticed there was almost no furniture in the apartment, and for the first time in memory, no weed in sight.

"Stephanie!" he called out. His young wife poked her head out of the bedroom. "Get James Junior here sumptin' to eat," he said, returning to his single-minded scrutiny of the pipe.

Stephanie led me into the kitchen and peered into the fridge. "Look inside this empty-ass icebox," she sighed, "and you pretty much lookin' at my life. If I don't get to his check before he do, there'd be no food in here at all."

Now that he was retired, Daddy got a pension check at the beginning and the end of each month. Stephanie told me he'd gotten a check three days earlier and smoked the last of it that morning. I sat at the kitchen table and ate a piece of bread with jelly smeared on it, just to be polite, while Stephanie explained how bad things had got. She still had her night-nurse job at the hospital, but she was afraid to leave her three-year-old daughter, Jyra, alone in the house with the sketchy crew he had hanging around.

"He smokes right in front of his own chile," she said. "He got no shame left." She told me he'd sold or pawned everything of value, even his old fishing and hunting gear. There wasn't a long gun or a shotgun left in the house. Stephanie was trying to persuade him to come with her and Jyra to Atlanta, where she had folks. Maybe they could get a fresh start there.

"James Junior," Daddy said in a hoarse whisper. I turned to see him in the kitchen doorway, cradling his big ol' .44 Magnum in both hands. "I need ya to hock this for me over at Kreeger Pawn. Git a hundred and ten for it—that's what dey gave me last time. And don't lose the ticket." That gun was his pride and joy, and I could see it pained him to hand it over to me.

"Then, I need ya to go over here"—he handed me a scrap of paper with an address scrawled on it—"and get me some rocks. Don't buy from anyone but Jimmy. Make sure he know it for James Plummer, y'hear?"

"I hear you, Daddy."

But all I could really hear was the sound of my childhood idol crashing off his pedestal.

WHEN I GOT BACK an hour later, the Kelly brothers were gone and Daddy was waiting for me at the front door. He took the rocks and went straight back to the bathroom to smoke. But he didn't close the door. I waited for him to come out, because I'd brought along my Tougaloo diploma to show him. Though Mama and Bridgette would eventually earn their high school GEDs, I was the first in our immediate family to graduate from college, or even high school. I figured it would make him proud to see my diploma, and to hear about Stanford.

After about ten minutes of watching him smoke through the bathroom door, I figured if I wanted to talk to him, I'd have to join him in there. Just as I stepped inside, Stephanie came down the hallway holding Jyra by the hand and kicked the door closed behind me.

It was tight in there and the white vapor tickled my nose and my brain. While Daddy hit the pipe, I explained how I'd gotten into four schools and had chosen Stanford because they offered me the most money. He nodded slowly up and down, but I couldn't tell if he was listening.

"Why'ncha go to LSU?" he asked between hits. "LSU a fine school. You don' need to go clear across the country." His voice was trembly, and his teeth were grinding against each other.

"LSU is good, Daddy. But Stanford is the best. And they graduate more Black physics PhDs than anywhere else—so that's my best shot. Plus, they're paying me to go study there."

"That's fine, then," he said, drawing deeply on the pipe. "You a grown man now. Uh-huh." When he opened his eyes they were fully dilated, and he looked straight through me.

"You wanna hit?" he asked, lifting the glass pipe in my direction.

"No, I'm good."

He put another rock in the pipe and hit it.

"It's getting dark, Daddy, and I left my car out front on the street."

He nodded. "You best be gettin' on then, I s'pose."

"All right, Daddy. See you later."

I got into my car and held on tight to the steering wheel to make my hands stop trembling. I realized I never did show Daddy my diploma.

PART FOUR

STANFORD STARMAN

If learning the truth is the scientist's goal, then he must make himself an enemy of all that he reads. He should also suspect himself, so that he may avoid falling into either prejudice or leniency.

—Ibn Al-Haytham (1030 CE), mathematician, astronomer, and inventor of the scientific method

53

IT DIDN'T TAKE ME LONG to figure out that I'd be starting off at the bottom of the heap at Stanford.

The first day of the fall term, the twenty new graduate students gathered in the lobby of the physics building. Someone passed around a sheet listing all our names alongside our undergraduate institutions. It read like a top-ten list of science centers from the United States and around the world: Harvard, MIT, Caltech, Princeton, Cambridge, Oxford, the Technical University of Munich, and the Moscow Institute of Physics and Technology.

The students seemed friendly enough, but I could feel their pretentiousness crowding me into the far corner of the lobby. I politely greeted a few students, but mostly just hung back and took in the scene.

The faculty member who presided over our gathering repeated what we'd all been told during our campus visits. Stanford selected us because we were the best applicants from around the world. I could tell that the other students were all quite confident in their abilities. And so was I, having just completed my first serious physics research project over the summer. I'd interned at the Center for Particle Astrophysics at UC Berkeley, helping a research team build a cryogenic dark-matter detector. I completed my project—engineering electronic antialiasing filters—far ahead of schedule. That same task had been assigned to a Berkeley undergraduate the preceding academic year, and he hadn't been able to build a single filter. So I was feeling pretty cocky.

As the students perused the list of names and undergraduate affiliations, they eventually became curious about the one school none of them had heard of: Tougaloo College.

"Who is James Plummer?" one wondered aloud.

"He must be a genius," another student joked.

In due course, they deduced that the shy Black guy conversing with no one must be the Tougaloo grad. Soon, I was surrounded by a small group of students who began questioning me—which is when I had my first run-in with a "Well, actually" guy, the physics nerd who can't help but correct statements made in error, no matter how slight or insignificant.

"What type of research do you do?" one of them asked.

"This summer I worked on a cryogenic dark-matter detector at Berkeley," I said.

Their expressions registered approval. I should have stopped there. Of course, I didn't.

"We used a disk of germanium to make direct detections of dark matter," I explained.

"By what method?" another student asked. "Dark matter is like weakly interacting massive particles, or WIMPs, so direct measurements are problematic."

"We detected collisions of the dark-matter particles with the germanium atoms," I answered, sensing I was already in trouble.

"*Well, actually,*" another jumped in. "germanium is a semiconductor, so any WIMP collisions would create electron-hole pairs and phonons that are read out by SQUIDs and FETs, respectively."

I didn't even know what a phonon, SQUID, or FET was, but I tried not to let it show on my face.

"Why did you choose Stanford?" someone else asked me.

"Because they accepted me!" I half-joked in response. My joke fell flat, so I quickly attempted to save face. "I wanted to come to the West Coast, and there's only two top physics schools, Stanford and Berkeley. I got accepted to both." This was a stretch. Berkeley initially rejected me, but offered me a spot in their grad program after my summer research with them.

"*Well, actually,* there are many good physics schools on the West Coast," a student retorted, earning nods of approval from several others. "In addition to Caltech, there's University of Washington, UC Santa Barbara, UCLA, UC San Diego . . ."

"Oh, I didn't know those," I replied, which drew heavy side-eyes from every direction. I felt deeply relieved when the "Well, actually" crowd lost interest in me and began chatting among themselves.

It was clear that I was the odd man out among my fellow grad students. Every year, the Stanford physics department took in one student like me—a diversity admission who wasn't at the same level of academic preparation as the rest of the class. It wasn't necessarily a Black student. It could be a Latino, or a woman, or even a white guy from a poor part of the country. I was the exotic and endangered species of wannabe physicist in my class.

My academic advisor, Professor Walter Meyerhof, explained how they planned to help me make up for having attended a series of schools in "underserved communities," as he referred to them. He recommended I take a year of senior-level undergraduate physics rather than diving straight into the graduate courses. I could tell he was worried that I'd be offended by the suggestion. But I was no fool. After spending the summer working alongside graduate physics students at Berkeley, I knew how much ground I needed to make up. It wasn't just the coursework I lacked. It seemed everyone else had been to Europe, had studied a second or third language, and knew how to order off a Japanese menu. So instead of protesting Meyerhof's recommendation, I volunteered to do *two* years of undergraduate classes. I calculated I'd be in my early thirties by the time I got my PhD. But so long as they paid me to study and do research, I wanted the job.

Meyerhof liked that I was ready to put my ego aside and do what it took to succeed at Stanford. "I'll be honest with you," he explained. "Things are changing in the department. I'm sure you're aware that Stanford has produced thirty African American physics PhDs in the last decade—more than any other school. Unfortunately, some of our younger faculty don't support this program. One of them is now the department chair. He was not in favor of your admission, but he was outvoted. That's why it's important that you do well in your graduate-level physics courses. If you get mostly A's in the grad courses and pass the qualifying exam, you'll have proven yourself. Passing the qualifying exam is the best way to tell my colleagues who don't support underserved students to back off."

The mention of qualifying exams brought to mind Miss Jolly's opinion that I wasn't qualified to advance from janitor to bellhop. I thought of the department chair, an atomic physicist who was a heavy favorite to win the next Nobel Prize, and wondered what it would take for me to convince him to someday accept me as a colleague.

I knew I was going to have to work in overdrive to compete with this Ivy League crowd. But I wasn't worried. I may have been the least academically prepared person in my program, but I knew I could outwork anyone. I'd carried a tuba through sweltering Mississippi summers, survived navy boot camp and Kappa Alpha Psi hazing. When it came to hard work, I knew I could handle whatever this country-club school could throw at me.

But I didn't know what I didn't know.

54

To be on the safe side, I decided to start out with three courses I'd already taken at Tougaloo. Only, they weren't the same courses at all. They had the same names and covered the same material, at least for the first few weeks. But the Stanford professors taught the courses at a much higher level, with much higher expectations of their students. They pushed the pace much more quickly and did much less actual teaching. They definitely considered themselves the smartest dudes in the joint, and it was our job to figure out how to operate on their level. As far as I could see, they saw their job as giving you a chance to watch them be brilliant up close—not to actually teach you the material.

I'd never seen anything like the problem sets they assigned. Every physics and math textbook had end-of-chapter problems that were typically assigned as homework. The most challenging problems were labeled with an asterisk or exclamation point. At Tougaloo, the professors never assigned the hardest problems. At Stanford, that's all they assigned. And when there weren't enough challenging problems to fill out a complete set, the professors would compose even more difficult ones for us to solve.

On the first day of each class, the professors issued the same warning: "I strongly encourage you to form groups among yourselves and work with others to complete the problem sets. Those of you who don't are unlikely to do well in the course."

I'd heard that the undergraduate physics students gathered informally at night on the fourth floor of the Varian Physics Building to work on problem sets together. So I showed up at Varian after dinner and camped out at an unoccupied table.

It turned out I was ahead of many of the students in my quantum mechanics class. I'd learned Dirac "bra-ket" notation and Dirac delta functions working one-on-one with Dr. Teal at Tougaloo. The other students saw that I was busting out those problem sets without much difficulty and several asked me how to utilize the delta functions. I was happy to help them.

But the next week's problem set on quantum tunneling, which relied heavily on linear algebra, had me stumped. After helping so many other students in the preceding week, I felt it was time to lean on them for a change. I approached a student and asked for help. He told me he couldn't help me. So I approached another and got the same response. I assumed they were all having the same difficulty I was. After working on it for another couple of hours, I got up to stretch my legs. That's when I noticed a group of my classmates working together and saw their papers on the table. They had completed most of the problems! Relieved that assistance was within sight, I asked them if they'd made progress on the problems and could offer me any help.

"Sorry," one of them said, while the others just looked at me. "I can't help you."

"Y'all haven't been able to solve the problems?" I asked. "I thought I saw that you did."

"We made some progress. But we're not ready to share yet. Sorry."

I then walked over to another group of students from the class. They turned over their completed solutions to the problems as I approached.

"Hi," I greeted them. "I'm having a lot of trouble with this problem set. Y'all know how to solve them?"

"Yeah," one of them replied. "We've solved most of them. There's a trick to it. Try working on it some more. You'll get it."

I was perplexed. I'd helped these guys on previous problem sets in any way they requested. And now they were straight up refusing to help me? One of the students from the class worked behind the checkout desk of the library near the door. Forget these guys, I said to myself. I'll ask him.

"Hi," I said. "I'm having a hard time with this week's problem set. The other guys told me there was a trick to solving these problems. I can get it to six equations and six unknowns—then I just drown in the algebra. I'm trying to plug and chug but it's just too much."

This guy decided to loud-talk me. "There is a trick," he said, so everyone could hear. "But you're gonna have to find it on your own. No one's doing it for you. *You* are responsible for doing your work, not us." Then, as I was walking away, he called out to me, "*Some* of us read before we ask others for help."

I stood there dumbfounded. I wanted to grab him by the face and kick his ass, or at least tell him off. But there was no way I was going to let them play me into being that guy. I packed my things up and left before I did or said something I would regret. I felt humiliated.

Fuck 'em, I thought. If that's how it is, that's how it is. I'll outwork all these muthafuckas.

I tuned my human dark-matter detector to high. From that point forward I resolved to go it alone academically.

55

CONTINUED TO WORK my butt off. My first-year graduate-student classmates soon began to joke that I lived in the physics library, as I spent virtually every night there. The only other places I saw grad students regularly was at the weekly physics colloquia, where top physicists from all over the world were invited to present their work, and in the one graduate physics class I took that first quarter: Research Activities at Stanford.

Most of it was a snore. I didn't see myself spending my whole career indoors, hunched over a lab bench or computer keyboard studying condensed matter or optics. I wanted to conduct some fabulous astrophysical experiment—like Eddington's 1919 eclipse expedition to measure whether light, which has no mass, is affected by gravity as predicted by Einstein's general theory of relativity. Or build something I'd launch into space that could peer into the far corners of the galaxy. I was turned on by the weird anomalies of the universe—the wormholes and the warped spacetime continuum. The only fields that truly captured my imagination were relativity, quantum physics, and astrophysics.

In 1991, Stanford wasn't conducting much experimental space research. Thank the Lord, Art Walker showed up with his soft x-ray photos of the sun.

At first, his presentation was as staid as any other physics professor's. He talked about the "coronal heating problem" and the "solar wind problem," and all the other problems scientists have encountered over the centuries of trying to observe and understand the star at the center of our solar system.

But once the slide show began, Professor Walker was like a proud dad showing off pictures of his kids. Except his kids were a rocket-mounted array of telescopes and his half dozen grad student researchers.

When the lights dimmed, an image of a black sun appeared on the screen. That's what it looked like to me. A large black disk with wispy rays of light emanating from around its edges. "This is an image of a total solar eclipse from March 18, 1988," Walker explained. "The black surface in the center of this image is the moon blocking the sun's surface from our line of sight, while allowing us to see the much dimmer white-light corona."

Until recently, he explained, the only way we could observe and photograph the sun's million-degree outer atmosphere, known as the corona, was during a solar eclipse, and all we could observe was the part of the corona showing outside the edges of the moon. But because the moon appears slightly bigger than the sun during a total solar eclipse, we can't observe the corona at its source on the sun's surface.

"To capture simultaneous high-resolution film images of the sun's full-disk corona," he explained, "we had to develop a new telescope technology: extreme ultraviolet and soft x-ray multilayer optics. My students and I designed and tested these EUV mirrors and filters, mounted them into fourteen individual telescopes, and launched them into space above the Earth's atmosphere, which would otherwise absorb the light we sought to capture on film. On May 13 of this year, our team launched the Multi-Spectral Solar Telescope Array, or MSSTA, from White Sands Missile Range in New Mexico."

The next series of slides showed Walker's grad students testing optics and working on the MSSTA payload at his Stanford lab. Then there were shots of the team at White Sands Missile Range preparing the MSSTA for launch. Watching Walker's students attach the payload to the rocket and prepare it for launch, I could feel the excitement build in the room. Images of the rocket lifting skyward off the launchpad gave way to photographs of Walker's team recovering the telescopes and cameras after the payload's brief flight in space.

"You will now be the first people outside of my research group to see the photographic results of our mission."

The first image he projected looked like a black sphere covered in white flames. "This is an image of the sun at 1.5 million degrees Kelvin as seen in the emission of iron 12."

I couldn't help raising my hand in the back of the darkened room and calling out, "Excuse me, sir. Did you say that's *the sun*?"

"Indeed it is," Walker responded. "The white loop structures you see at mid-latitudes, the plumes emerging from the poles, and the compact bright points distributed all over the disk are solar coronal structures. This is the highest-resolution image of the sun's extreme ultraviolet corona ever obtained."

It was an astonishing sight. Gasps of surprise and wows of appreciation filled the room as image after fiery image revamped everything we thought we knew about the appearance of the sun. My mind raced, trying to assimilate the implications of those images: So that's what the corona of the sun *really* looks like? Whoa! What must that mean about the hundreds of billions of stars in our galaxy and the trillions beyond? Were they all "aflame" like that? I imagined the universe as a matrix of stellar bodies, each a massive ball of super-hot matter covered in a giant forest of even hotter plasma-filled magnetic fields that convulsed with tempestuous storms of winds and plumes and sunspots. I would never think of the sun or the stars the same way again.

In the midst of all the hubbub his photos aroused, Walker smiled delightedly right along with us at the magical images he and his team had captured. It was as if he had handed us all 3-D glasses and we were seeing the stars in multiple dimensions for the very first time. Sure, Dr. Walker was what I'd describe as "whitewashed," and yes, he was part of the solar physics elite—but there was something pure and almost childlike in his enjoyment of the stellar world he was exploring.

Standing there in the back of the auditorium, I realized I was now one of a select few humans anywhere on Earth to behold the invisible face of the sun. After tens of thousands of years of squinting up at the bright orb at the center of our solar system and wondering, we could finally see the sun's true nature.

I didn't yet understand the images' significance, or why they were important to our understanding of other stars in other galaxies. But it felt as though those photographs opened up the top of my skull and stirred up everything inside. Just then, I didn't care about the race-baiting and the preppy snobbism and the put-downs I'd endured during my first months on campus. All I knew was that I wanted in on Art Walker's research group.

56

B Y THE END of the first term, I was screwing up any chance I had of joining Art Walker's, or any other, research group. I was working as hard as I could, but I was working on my own. And that wasn't gonna get it done. I got B's and C's in courses I'd gotten A's and B's in at Tougaloo. That felt like a colossal fail. Usually, I worked my butt off when I had to, and then there'd be a win and a big emotional release. This time I'd worked harder than ever—and still took a massive loss.

Things went downhill even faster at the beginning of the next term. I was taking totally unfamiliar math and physics courses, and I was still trying to go it alone. When we took our midterm exams, I got the lowest score in each class.

I was falling through the floorboards, and there was no one to catch me. If I quit, I'd have nothing. I couldn't go back to Piney Woods, and I couldn't see how to keep up at Stanford. One night I called Bridgette in tears. I hadn't cried in front of her since I was ten years old.

"Why you callin' and cryin' on the phone to me?" she asked. "I'm not your mama." I didn't know where Mama was exactly—she was in between home addresses at the time—and I certainly didn't want to admit to her that I'd climbed up too high in a big old oak tree and didn't know how to get down. Meanwhile, Bridgette had her hands full raising up her own child without me, a grown man, coming to her with my boo-hoo face on.

I DECIDED IT WAS TIME to go see my academic advisor, Professor Meyerhof. Maybe he could tell me how to survive my first year. He wasn't in his office, so I sat on the floor with my back against the door of his office and waited for him to return.

After half an hour, Professor Doug Osheroff came out of his lab across the hall. I hadn't seen him since a couple of months earlier, when I had a brief and unsuccessful research rotation in his super-fluid physics lab. He'd given me an assignment that was light-years over my head—something having to do with measuring properties of superfluid helium-3, a phase of helium he discovered for which he'd win the Nobel a few years later. I was too embarrassed to re-turn to his lab meetings without performing my assigned task, so I'd been artfully dodging him ever since.

Crouched on the floor outside Meyerhof's office, I imagined drawing a cloak of invisibility around my shameful condition.

"Hey, James," he called out cheerfully. "What're you doing sitting on the floor?"

"I'm waiting to see Dr. Meyerhof."

"Oh, yeah? What about?"

I mumbled something about academic challenges, and maybe dropping out of the program.

"Why don't you come in and talk to me about it," he said, motion-ing me to follow him into his office. He didn't mention the assign-ment I hadn't completed or the book on superfluid helium he lent me that I never returned.

"Why are you thinking of dropping out?" he asked me.

"I don't think I'm smart enough to make it here."

"What makes you feel that way? We thought you were pretty smart when we admitted you."

"I just got my midterm exams back and I got the lowest score in each class," I said.

"I see," Osheroff replied. "What courses are you taking?"

"Electricity and Magnetism, Quantum Mechanics, and Partial Differential Equations," I answered.

"Okay," he responded. "What are you tested on in Quantum?"

"We covered the one-dimensional harmonic oscillator and the hydrogen atom," I replied.

"Very well then." He handed me a piece of chalk and motioned toward his blackboard. "Can you write down the time-independent eigenvalue equation using the quantum mechanical Hamiltonian for a one-dimensional oscillator?"

"Sure," I responded, and wrote the equation on the board.

"Okay. Now, using the raising and lowering operators and their commutator relation, can you derive the energy eigenvalues?"

"I think so," I replied, and began working. In short order, I derived the energy eigenvalues he requested.

"Great," Osheroff said. "What was on your PDE midterm?"

"We worked on the heat equation in three dimensions."

"Okay. Write down the heat current density in terms of the conductivity and temperature gradient."

I again followed his instructions without requiring help.

"Not bad," Osheroff said. "Can you derive the heat equation using the divergence theorem?"

"Yes," I replied, and worked through the math on the board.

"How about one more?" Osheroff asked.

"Okay."

"I'll tell you a geometry and boundary conditions and you calculate the separated solutions to the heat equation."

"Okay." And for a third time, I solved the problem as he requested.

"I don't know, James," Osheroff said. "Those are some tough problems and you solved all of them with no sweat. You sure seem pretty smart to me."

"Maybe. But *everyone* performed better than I did on those exams."

"There's a difference between performance and mastery. Performing badly in your courses for one academic quarter is not going to derail your career. Remember that you're working toward mastery of physics and mathematics, which takes a long time. This is only your second quarter here, and you're competing against students who've had every academic advantage since kindergarten. But you have the smarts and work ethic to make it. Can you promise me that you'll stick it out for the remainder of the year? After that we can talk about it again."

I promised Osheroff I'd stay. I didn't stick around to talk to Meyerhof, since Osheroff had told me what I'd been hoping to hear. He reminded me that my smarts and hard work were what got me there. What I needed now was tenacity and fortitude, which for me meant putting up with the petty humiliations of my classmates and the humbling experience of having to work my way up from the bottom.

57

WHILE I WAS teetering on the edge academically, Jessica headed back to Jackson for her final semester of nursing school. When she was packing the night before she left, I heard her singing Gladys Knight's "Midnight Train to Georgia": *"She's goin' back to find . . . the world she left behind . . . a simpler place and time."*

Ever since we'd arrived at Stanford in late August, Jessica had been feeling like she didn't belong. At first, grad school had felt to us like a game-show giveaway. Not only was Stanford paying me $15,000 a year to study and do research, they also gave us a furnished apartment in a graduate housing complex called Escondido Village. The week after we moved in, Jessica landed a job at a sandwich shop on campus. The woman hiring her apologized for only being able to pay Jessica $7 an hour. We laughed our asses off. Seven dollars an hour was big money to us.

But Jessica never really felt at home. Not in Escondido Village. Not on campus. Not in Palo Alto. The first time we pulled into Town & Country Village, the upscale shopping center on El Camino Real, and parked our little beat-up Nissan Sentra alongside a row of gleaming new Beamers and Benzes, Jessica leaned her forehead against the dash. "We don't fit in here," she groaned. She felt the same way on campus when we walked across the perfect green lawns past all the country-club girls and boys throwing Frisbees and lying out on towels. She missed hearing Southern voices and eating Southern food. We both did.

When we got good and tired of hanging around white folks all day, we drove east across the Caltrain tracks to East Palo Alto—the Black side of town. EPA was the shadow alter ego of Palo Alto. Three miles from Stanford, just east of Route 101, EPA had the highest per capita homicide rate in the country that year. Crack had taken the community by storm, its drug gangs out-murdering even those in the ghettos of Compton and Oakland. The McDonald's we went to at University Avenue and Bay Road had bulletproof glass that was pockmarked from drive-by shootings.

That corner epitomized EPA's history. It was previously a small commercial area that included a Bank of America and a supermarket. Across the street was an Afrocentric Black community known as Nairobi Village, since its inhabitants had all learned Swahili. It wasn't unusual to spot Black celebrities there. Muhammad Ali, Shirley Chisholm, and the R&B group the Whispers would regularly stop by. But now EPA had no bank and no supermarket. And it had gotten too damn dangerous for celebrity drop-bys.

Still, I felt more at home and less on guard in EPA than I did on Stanford's campus among the "Well, actually" set, where I had to remember to talk like a proper science nerd and never drop my consonants. EPA's streets reminded me of Black neighborhoods in Jackson, Houston, and South Central L.A. My dark vision could see the dirt on every corner of EPA, and there was plenty to go around. But there was also a taste of home there. Rocky's BBQ served up ribs, baked beans, and cornbread that tasted just like Piney Woods. Across the street was West Sounds record shop, where I could grab some incense and rap cassettes from Ice Cube and Boogie Down Productions. I could pick up some urban-looking threads at Style Setter. Chet's sold every Black magazine, as well the full spectrum of smoking paraphernalia. Across town at the Gardens, I could get a fresh cut at Walker & Price's barbershop and some chicken and waffles at Crystal's Diner. At the border of EPA on our way home, we could score Afrocentric books at Markstyle Bookstore.

Then all of a sudden we'd be back in Town & Country Village with its Olive Tree Mediterranean food court and no-soul Muzak.

ONE OF STANFORD'S saving graces was the feeling of community among the Black students. We had our so-called safe places. There was the Black Community Services Center (aka "The Black House"), where folks would congregate between classes. There was even an African American–themed dorm called Ujamaa House, where mostly undergrads hung out in the evening. Undergrads and grad students also had their own organizations: the Black Student Union and the Black Graduate Students Association.

The true Black undergraduate elites at Stanford were the Division I athletes, the football and basketball players hoping for pro careers, and the other sports stars laser-focused on making the U.S. Olympic team. Being a Division I athlete was a full-time job, which meant most of them didn't hang out drinking or smoking weed, or really partying much at all. Tiger Woods didn't arrive until my third year at Stanford, but he was cast from that same mold. All business, he went pro after his sophomore year and joined the PGA Tour.

As was the case in every Black community I'd ever lived in, the Black women were serious and in charge. The undergraduate women were plenty fun, but the graduate student women were all business. They were cool, but they were serious cool, letting you know they planned to be leaders in their fields. The Black men were playing catch-up. The attrition rate among the undergraduate Black men was a known problem, even though many of them had attended prominent prep schools like Phillips Exeter Academy in New Hampshire and the McCallie School in Chattanooga, Tennessee. Finally, there was a small number of Black students who were "blinded by the white," who behaved as if they just wanted to disappear into the white, preppy heaven of Stanford. They joined the white fraternities and sororities, and only dated interracially, wanting nothing to do with the Black student community at all.

Luckily, Stanford had a Kappa Alpha Psi chapter, though unlike the white fraternities on campus, the Black frats didn't have dedicated houses. My first week of classes I ran into a brother wearing a Kappa shirt. I hit him with our secret signal, and he invited me to meet with the local chapter that weekend. They were altogether unlike Southern Kappas—and they found me as weird as I found them. Most of the brothers had absolutely zero hood in them. And none of them had any Southern. When they said "down south," they meant L.A.

ONE OF THE FEW Kappa brothers with some hood was Troy. He'd grown up on the streets of Oakland before going east to Dartmouth. After college, he came back to the Bay Area, got an IT job in Menlo Park, and then found a management gig at Stanford's Mail Services. He lived with a couple of other Kappa alums in a house off campus. Troy was super charismatic. He could sing his ass off, and women considered him a true "pretty boy," which he was. When it came to being well groomed and well dressed, Troy was in a class by himself. He was the first Black man I'd ever met who got pedicures. And like me, Troy loved to smoke weed.

After Jessica left for her spring semester at Jackson, I found myself dropping by Troy's place almost every night after the library closed. We'd chat and smoke in his kitchen, and he'd down shots of Stoli vodka that he kept in his freezer.

For me, hanging with Troy was a welcome change of pace from the square white nerds in my program. It felt totally chill, even when the coke first showed up.

58

IT TURNED OUT that Troy liked to snort coke on weekends, or else sprinkle it on a joint and smoke it as a "primo." Being a son of Oakland, he knew his way around EPA, where he had a connect—an old girlfriend of his named Betty. She was from Mississippi and had come up from the ghetto like Troy and me. After graduating Stanford undergrad, she'd landed a good job in administration at Stanford Medical Center.

We'd score a gram or two from Betty, and then we'd hang at her place and do lines or smoke primos while we played cards. I quickly felt at home around Betty, with her down-home accent and her dynamic, bossy personality backed up by a brick-house body. Growing up with my mama, I knew and respected her brand of strong, smart, and assertive Southern woman.

I'd never been into snorting coke. And once you've smoked rocks, smoking a primo is simply frustrating. Smoking powder didn't hit hard like a rock did, didn't give you that nice "Boom!" Once you've tasted the real thing, powder doesn't quite get you there anymore.

One night, while I was watching Troy lay out lines on his kitchen counter—which to me was a waste of good coke—I said, "Hey man, do you ever think about rockin' it up?"

"What do you mean?"

And I said, "Let me show you."

I took him back to my apartment in Escondido Village, emptied a small glass jar from a kitchen cabinet, added some water and baking soda, and showed him how to cook rocks from powder on my stove. I didn't have a pipe, so I punched a circle of pinholes in a soda can and lit one of Troy's cigarettes. I placed some ash over the pinholes, placed a rock on the ash and showed him how to melt the rock into the ashes and then fire it up.

Troy had never smoked rocks before. From a selfish point of view—and that's the only point of view there is when crack is involved—I was hoping he'd like it. Sure enough, his eyes went all dilated and glassy and his face was split in half with a grin so wide you'd think he just discovered sex or something. I couldn't wait to get my turn on the can.

The first time you smoke a rock after a long time away, you're right back where you were the last time you did it. At first, you get good and high, like the best kind of welcome-home party. But pretty soon, one thought overpowers your mind: *We must do this again.* When the rocks run out, you get super sad. So naturally, you have to go back out and get more.

You would have thought there'd be alarm bells going off in my head, warning me away from the rocks, or even from powder. I'd seen firsthand how crack could enslave you and steal your soul. It had taken down JG and my daddy, and it almost took me down. And yet, here I was in Escondido Village—a few hundred yards from the physics library—showing Troy how to cook and hit that can. The next day, walking into Chet's Smoke Shop in EPA and buying a glass pipe felt like the most natural move in the world. It affirmed the part of me that I never stopped believing in. The dark side.

As soon as we started cooking back at my place, Troy and I talked about how we had to keep this secret—from everyone on campus, and definitely from Betty. Which might sound funny, since she was our main connect. But Betty was a no-nonsense-type chick, the kind of kick-yo-ass matriarch we both grew up with. She knew we had real lives to tend to, and we figured she'd come down on us heavy if she heard tell of us smoking rocks.

It wasn't long before we were scoring eight-balls instead of a gram or two from Betty. And we weren't hanging out and partying with her anymore. One night we showed up multiple times at her apartment in EPA to reboot. Betty had a serious job she had to get up for in the morning. So the third time we showed up at her door, she got good and cranky. "There's no way you guys are snorting this much powder," she said. "Are you cookin' it?"

That put us on edge. Since we were too tweaked out by that point to credibly deny we'd been cooking, we figured we were about to get our asses properly kicked. Instead, Betty gave us her widest smile and said, "Well, shit. Let's cook some now."

ONCE BETTY CAME ABOARD, the party train really left the station. Betty had a friend who rounded out our crew. Theresa was a strikingly beautiful, light-skinned girl with thick, rich hair—one of those chicks who wore her jeans low with a thong showing up above. Theresa had gone to Palo Alto High, which was a Northern California variation on *Beverly Hills, 90210*. She was an innocent girl who still lived with her mom. But wherever Betty went, Theresa was sure to follow. Troy, Betty, and I were aspirational professionals who liked to party on the side. Theresa just liked to party.

None of us were on Easy Street, financially. So we could only afford to party a few nights a week. And since most of us had lives we had to show up for daily, I wasn't really worried about going off the deep end. After all, Troy wasn't JG.

But there were plenty of red flags. When Betty's connects ran dry late at night, we ended up visiting "spots" Troy knew about where we could score. The minute you pulled up, your car was surrounded by half a dozen roughnecks. Most were there to sell, but you had to watch out for the ones running a game. They'd push baggies through the top of your window as an excuse to get you to open your door—at which point they'd drag you out, shove a gun in your face, and rob you. The trick was to never let your window down too far, never open your door, and never show fear. If those dudes smelled fear on you, you might just as well hand over your cash *and* the keys to your car.

As the spring wore on, I went in hard and heavy. Harder and heavier than Troy had the stomach for. After one seriously evil session when we smoked up Troy's tax refund in a nonstop thirty-six-hour bender—and after our third run-in with the wrong end of a gun at some fool spot—Troy decided to save his life and take a job in L.A.

Theresa was the vulnerable one in our crew. Unlike the rest of us, she never got the knack for going hard for a night or two, then laying off for a week. Theresa became a hundred-percenter right off, and now she was heading straight down the tubes. I felt guilty about pulling Troy into the life, and worse about Theresa. Betty had turned her on to crack, not me. But I knew that we ghetto kids who'd come up from the street were made of tougher stuff than a multiracial beauty from Palo Alto High.

59

For a while, I was able to juggle my double life on the street and on campus. There was James the rock man, who slipped into EPA with his posse to score and party for a nonstop weekend. Then there was James the aspiring rocket man, whose academic performance actually improved that spring. I was acclimating to the higher academic standards of Stanford, becoming proficient in the language of physics in its highest and purest form. I lost all self-consciousness about being the beginner who repeatedly interrupted folks to say, "I'm sorry but I don't know that word." I was a fast learner. Once someone explained a word or term to me, I could rap with it like a pro.

By the end of the spring term, I'd been invited to work with Art Walker's group. I moved into one of the shared graduate student offices in the Electronic Research Lab and showed up regularly at our beamline experiments at the linear accelerator.

Earning a slot in Art's research group was no gimme. For starters, I had to survive first-year academics. I didn't get A's, but my year-end grades were respectable. Then, Art had to want me. When he tasked me with modeling how the MSSTA telescopes would interact with the light emitted by the sun's corona, I aced it. That research also laid the groundwork for my first scientific publication.

Before I could sign on, his research group had to check me out to make sure I'd be a good fit. Art's group had a reputation for being extra-cool physicists. They were all whip-smart nerds, of course, but they had the only student office with a serious stereo. Craig DeForest was a brilliant long-haired hippie who drove a motorcycle and never wore shoes. His two office mates were Ray O'Neal, a MacGyver-style problem-solver, and Charles Kankelborg, a born-again brainiac who made a hobby of picking locks. Adding a cool nerd from the hood to their crew made sense to them—so I was in.

MY BALANCING ACT got trickier when Jessica returned home in June. I had to cover my tracks at night, saying I was working late at the lab—which sometimes I was, and sometimes I wasn't.

In July, Jessica got pregnant.

You'd think the prospect of becoming a father again would have slowed me down. But it brought back all the super-sad stuff with Lisa and Martel, and the deep, dark rabbit hole all that sent me down.

Things hadn't gotten any better for Martel. After Lisa moved to Houston, she left him with Miss Banny in Wesley Chapel. I was paying child support, and Social Security was paying for his therapy. But Miss Banny wasn't taking Martel to his therapy sessions in Jackson. When I was still at Tougaloo, I would drive to Wesley Chapel, pick Martel up at Miss Banny's, drive him to Jackson for therapy, and then drive him back to Wesley Chapel. But I couldn't do that regularly, since I was still in school. The therapist in Jackson explained to Jessica and me that if Martel could acquire certain abilities before he turned seven, like walking and talking, he'd have them for life. But since he wasn't getting to therapy much, he was falling further behind. When Jessica and I tried and failed to bring him to Jackson to live with us, I felt like a miserable failure as a father.

Right after we arrived in Palo Alto, Jessica began checking out the pediatric nursing opportunities at Stanford Medical Center, since that was her specialty area. She found out they had a center for cerebral palsy that seemed like a perfect fit for Martel. So once again we tried to bring him to live with us. Miss Banny told me, "That's Lisa business." But Lisa wouldn't even talk to me about letting Martel come to live with us.

"What's the problem?" I asked her face-to-face when I was back in Wesley Chapel for Christmas during my first fall in grad school. She didn't respond. "Is it that you don't want another woman raising your child?" Lisa just crossed her arms and stared at me without saying a word. That woman was made of stone. And it broke my heart.

MY SHAMEFUL SECRET LIFE on the streets—a familiar shadow that had come back to haunt me—made it hard to feel good about Jessica and me having our own baby together. It felt like the secret was eating all our happiness, and I didn't know how to tell her about it. About how I'd been chasing the rocks through the streets of Jackson when we were dating at Tougaloo—and how the rocks were back now, calling me across the tracks to EPA and away from our home.

I didn't have any coursework that summer, and with Troy gone, there was no one stepping on the brake pedal. Betty and Theresa and I were bingeing with abandon.

As soon as the fall term started up, things were stretched dangerously thin. I'd begun to TA Art's undergraduate astronomy class, and I had a full course load of my own. Plus, I was working with Art's research group at the Stanford Synchrotron Radiation Lightsource (SSRL), which I definitely didn't want to fuck up.

But the rocks kept calling, *C'mon and git me*. It began each evening as a singsong invitation. By midnight, it was a command.

60

EVEN THOUGH MY HEAD felt like it was about to go all supernova on me, Saturn's coordinates from the Observer's Handbook appeared in my mind's eye as if tattooed on my retina: "Right ascension: 20 hours, 59 minutes, 6.4 seconds," I intoned. "Declination: 18 degrees, 20 minutes, 33.2 seconds." As I dialed in the coordinates, I could hear the gears of the Boller & Chivens telescope mount grind into place. But when I looked through the eyepiece, a passing cloud was blocking my view.

My mind was somewhere else. I had a rendezvous scheduled for midnight, and the seconds were ticking down like one of those movie close-ups of a clock's second hand with the click-click-click soundtrack amped way up. I was counting down the seconds with one part of my brain while another part calculated how many tasks I needed to complete over the next 3,500 one-second intervals. That's how I got when the rocks started calling to me at night. My brain went all OCD timepiece on me, and the clockworks wouldn't wind down till it got fed.

It was my first term as TA for Art Walker's undergraduate Observational Astronomy class. Physics 50 was known as an easy A for fuzzies—the name we gave to humanities majors who needed to fulfill their one-semester science requirement. I was lucky to have this TA gig funding me through the end of the term. But I couldn't seem to unclench the fist inside my chest or muffle the low moaning sound, like some wounded animal, that called to me through a narrow tunnel in my head.

The cloud finally passed from view and the rings of Saturn rode to my rescue. Viewed with the naked eye across millions of miles, Saturn is just a dim yellow point of light in the sky. But magnified through the 14-inch Schmidt–Cassegrain scope, the rings always knocked my eyes back in my head. I gazed in wonder, as if for the first time, at the three major rings—170,000 miles of ice and rock held in frozen circumference around Saturn. The sight of the rings calmed my brain for the first time all night. Now that I had a clear view of them, I didn't want to move away from the eyepiece.

Then I remembered my midnight meet-up, and all the students who needed to view the rings before I could close up the observatory for the night. Reluctantly, I stepped away from the scope and motioned for the first student to take a look. One by one they oohed and aahed with broad grins of first-time discovery.

AT 11:28 I JUMPED on my bike and pedaled down the hill to the Stanford Synchrotron Radiation Lightsource. Our research group was using the linear accelerator to measure how well our telescope mirrors reflected the x-rays and extreme ultraviolet light as a function of wavelength. If our mirrors tested out, we'd install them on nineteen telescopes aboard the Multi-Spectral Solar Telescope Array II rocket set to launch the following fall.

I figured I could duck into the lab, check on the vacuum chamber, get a run going, and bounce back out by 11:55. I prayed that Art was home with his wife instead of checking on his night-shift students. Time on the beamline was very precious and competitively awarded, so experiments ran 24/7. Art's group was working twelve-hour and six-hour shifts. My shift didn't start till six A.M., but I wanted to check the helium cryo pump. When it shut down a few nights earlier, we'd lost almost a whole day getting it working again.

When I entered, Dennis Martinez was pouring liquid nitrogen into a row of sorption-pump dewars. He was splashing the shit everywhere, so I gave him a wide berth as I made my way over to the cryo pump.

"I'm almost done here," said Dennis, looking up from his vacuum unit. "It's gonna take a while to pump down again. Want to grab a cerveza?"

I got along okay with Dennis, the Ecuadorian third-year on our team. But my midnight witching hour was drawing closer with every passing click in my head.

"Sounds chill," I said, relieved to see my pump's ion gauge was holding steady at 10^{-7} torr. "But I promised Jessica I'd be home by midnight. Now that she's in her second trimester, she's hungry all the time and waits up to eat a second dinner with me. I'll catch you at the other end of your shift."

AT 11:56, I WAS back out in the cool night air, coasting downhill past the manicured lawns and night-lit fountains.

At two minutes before midnight, I rolled into Escondido Village and locked my bike in front of the entryway. But I didn't go inside. Instead, I made a hard left out of the quad and walked quickly down the slope to El Camino Real.

The sight of Betty's car idling at the curb delivered a potent cocktail of relief, excitement, and guilt. Once inside the car, I couldn't wait to get full. Betty had brought Theresa along, as usual. Our plan was to roll over to EPA to score, and then head back to Betty's place to party. But it was still a loosey-goosey plan, with Theresa going on about some "spot" she knew where we could score.

"Hell no! I don't go to no spots!" I shouted over the Geto Boys blasting from the car speakers. "Spots are where you get yo' ass shot!"

"Don't you worry, chile," cooed Betty. "I got us an *indoor* spot that's no-fail. Rocks around the clock!"

"Just don't take us to no street spots, Betty. I'm serious!" I said while Theresa laughed behind me.

I'd been down that road too many times to believe in "no-fail" spots. First off, it was always dicey showing up for a ghetto hookup with two fine-ass sistas in tow, particularly if they were dolled up like Betty and Theresa were that night. Second, I could tell that the ladies had already been hitting on something, so we were flying snow-blind before we even took off. Third, I'd had my gun stolen off me some months back, so we had no protection other than whatever street wits the Lord blessed us with—which didn't amount to squat, seeing as how we were rolling into EPA after midnight without a solid plan.

"Just so y'all know," I said, "I have to be back at the lab by six."

Theresa leaned over from the back seat, rubbed my head, and laughed her jaybird laugh. "Don't you worry about a thang, Rocket Man. We gonna set you right tonight." She lit a primo and passed it forward.

As we crossed Route 101, I promised myself we'd score enough dope so we wouldn't be roaming the hood again in a few hours, tweaked out and hitting spots on dark corners. We'd cop, go back to Betty's, and party in peace.

IT WAS 1:15 and the "rocks-around-the-clock" still hadn't appeared. We'd already spent an hour finding the spot, getting inside, getting upstairs, and waiting around for someone to actually show up with the dope. I'd already counted the blue cornflowers on the stained wallpaper twice—forty-three diagonal rows of sixteen flowers summed to 688 blue cornflowers, all in a latticed array, like atoms queued up in a semiconductor.

At 1:40, some tweaked-out dude in stained jeans and a ragged tee showed up with the goods. Theresa was off in the corner being chatted up by some unwholesome fake-pimp-looking muthafucka I suspected I'd have to detach her from before we could get out the door. I just wanted to take the rocks and bounce, but Betty insisted we verify the product. "We" meant me. Betty always wanted me to vouch for the shit.

I loaded one end of the straight shooter with a bollo and curled my lips around the other. I fired it up and sucked soft and steady, then hard. Final countdown to launch.

It took just seconds for the vapor to exit the dingy glass tube, pass through my throat, and enter my thirsty lungs. The smoke suffused the pulmonary capillaries, bonded with my blood, shot through my heart, and caught the express train to the blood-brain barrier impatiently awaiting its reward.

We had ignition. And lift-off.

Finally, mercifully, the clicking clock in my head shut off. The fist in my chest unclenched. The low animal moan was smothered into silence by the blizzard in my brain. As my eyelids drooped, the rings of Saturn returned to view. I could see them so clearly now, the numberless motes of dust and ice encircling my head like a frozen halo.

A TIMER WENT OFF in my brain, telling me it was time to split. But by then Theresa had gone off across the street with the fake pimp who probably promised her some bigger, better rocks. Betty left to go find Theresa, and I was alone with some woman I didn't know who pulled out a pipe and offered it to me. The sensible thing— going back home to Jessica—was the hardest thing to do right then.

An hour later when we were empty and the woman said, "Let's hit the ATM at 7-Eleven and keep it rolling," I was too desperate for more to say no.

I pulled my last hundred out of the ATM and we were walking back to her crib when a jacked-up Buick pulled up alongside us and four punks jumped out. They looked about fourteen years old, but were hard around the eyes. She called out to them, "Y'all want some money, get this muthafucka!"

I saw something flash silver, which I figured for a knife, but when it came up to eye level I saw it was a snub-nosed chrome-plated piece. Someone shoved me hard against a chain-link fence, and I felt the gunmetal cold against the back of my head.

I whipped off my wristwatch and waved it in the air. One of them snatched the watch while another rifled my pockets for the cash.

"Shoot his ass!" the woman shouted behind me.

I was too dazed to cry out. Too high to run or fight. My luck had run out. I'd known plenty of people who'd died one way or another in the trap. It was my time now. I was gonna die right here on this sidewalk. I'd never see Jessica again. I'd never meet our child.

I gripped the chain link with both hands and waited. I heard the gun cock and then the loud pop. They howled with laughter while I dropped to my knees and wondered why I hadn't felt the bullet enter my head. When I heard them jump into the car and peel away, I realized they'd shot into the air.

They were gone. The fear and the tears would come later. Right then, I felt nothing. I was dead empty.

61

Trudging up the hill at dawn toward the SSRL, I felt like Stanford's offensive line had used me as a blocking dummy all night.

Stripped of cash, I had to walk the three and a half miles back to campus. Without my watch, I didn't know the time. But I could tell from the position of Orion in the southwest sky that it was close to six A.M., when my shift at the lab would begin.

Once I crossed the Caltrain tracks, my numbness receded and my emotions came surging back. And they were all bad. I felt sick to my stomach and sick at heart. I hated myself for the hole I'd dug and dove into. And I was ashamed to have led my friends down that same hole. Ashamed for letting down Art and my research team and Jessica and our unborn baby.

I was tired and I was scared. Scared of dying. Scared of what I'd have to do to keep from dying. But I couldn't continue like this. Something had to fucking change. I had to talk to someone. But who? I couldn't go to Jessica. She had no clue that I'd ever been in the life, and I didn't want her to see me that way.

When I entered the lab, Dennis Martinez was sitting at the computer at our beamline, logging out after his shift. I must have looked pretty ragged because he scanned me up and down and said, "What kind of saloon have you been drinking in, hombre?"

Dennis was one of the older grad students, having come to physics as a Lockheed engineer. He'd been a diversity admission a couple of years ahead of me, so I guess that's why I unloaded in front of him. Or maybe I would have bared my soul to the first human being I encountered that morning.

"Listen," I said. "What I'm about to tell you, it's gotta be a dead secret. You gotta promise me you won't tell anyone."

Dennis didn't look too thrilled to be my confessor. He just shrugged and said, "Okay . . ."

So I told him. Not everything. But enough. I told him I'd been into some bad shit. Drugs and such. I spoke in a depressed monotone, and the words kept coming out. I told him I'd had a gun pulled on me, for the umpteenth time, and I thought I was gonna die. But I didn't know how to stop. And I didn't know what to do.

Finally, I stopped talking. He looked like he didn't know what to say to me. What he said, finally, was, "You gotta talk to Art."

Talking to Art was the last thing I wanted to do just then. I told Dennis, yeah, I should probably talk to Art, and made him promise again not to tell anyone what I'd said. I still had a six-hour shift ahead of me, which was a gruesome prospect with no sleep and nothing for comfort or company but the unforgiving fluorescent light and the vacuum pumps wheezing like a bunch of old geezers ready to croak.

At noon, when my shift ended, I had nothing else to do and nowhere else to go. It was time to face my mentor.

62

WHEN I FINALLY FOUND the courage to knock on Art's office door and venture inside, I found myself in a forest of stacked documents. From the doorway of his inner office, I could see him working at his desk, partially obscured by several tall stacks.

I knocked on the doorframe, but he didn't look up from the document he was highlighting and notating. After a minute, I knocked harder. He peered at me over the top of his glasses without lifting his head.

"Art, could I talk to you?"

"Okay," he said as he slowly removed his glasses and sat upright. "Have a seat."

There was nowhere to sit, so I had to clear a pile of papers off the chair in front of his desk. I'd tried without success to plan a speech on my walk over from the lab. This was the first time I'd been alone with Art in his office since our meeting when I was first accepted into the program. For the past eighteen months, I'd been looking for an excuse to get some face time with him and build up our relationship. That's what every grad student wants to do with his mentor. Now that I was finally alone in his office with him, I had to tell him that I'd let him down.

I knew Art might drop-kick me out of his office, and out of the program. But I almost didn't care. All I knew was that I was bone tired from carrying the heavy load of my double life around, and I needed to set it down. But I didn't know how or where to begin.

Art must have known something was wrong because my vibe was so down. I was usually the high-energy, excited-electron guy in his lab. But he didn't say anything. He just waited calmly with his fingers tented against each other.

"It started at Tougaloo," I began. And I told him everything. About Martel and my dad, about cooking and smoking rocks, about dropping out of school and hitting the streets with JG, and all the other nasty stuff I was doing back then. Then I told him about going back in this past winter—even though I wanted more than anything to succeed at Stanford and contribute to his team and be part of his next rocket launch. And finally, I told him about how last night had ended with a gun at my head, and how I couldn't continue with all that, how I knew I had to come clean with him if I wanted to get clean and restart my life.

The whole time I was talking, Art just sat there and listened to me without any expression on his face. When I ran out of things to say, he let everything I'd told him hang in the air.

"Well," he said finally, "you're not going to do those things anymore, are you?"

I thought I must have missed something. The flywheels in my mind were spinning madly to make sense of what he'd just said. Could it be that easy? Could I make that commitment to Art and keep it? Could he really believe in me, after everything I just told him?

"No," I said, "I'm not going to do those things anymore."

"Good. You're married, right?"

"Yes," I said. "We're expecting a baby."

"Does your wife know what you've been up to?"

"No, sir."

"Then you need to go talk to her."

"Yes . . . I will."

I knew that was my cue to get up and leave. But I couldn't seem to move.

Art must have sensed my inertia, because he pushed his chair back a bit from his desk and sighed. "You know, you're not the first Black man to hit some speed bumps on the way up. I grew up in Harlem, and my mama made it her business to get me into a fancy Jewish school in Morningside Heights, and then into the Bronx High School of Science. But that didn't mean teachers believed I could have a career in science. My chemistry teacher told me that if I was determined to be a scientist I should consider moving back to the islands where my father came from, or maybe Cuba. My mama just about ripped his head off." He smiled at the memory.

"That wasn't the first or the last time someone told me 'you can't.' You think Stanford is white? How many Black PhD students do you imagine there were in astrophysics at the University of Illinois in the fifties? The Air Force was happy to have me doing research in their weapons lab, but they weren't handing out leadership positions to men who looked like me. I had to put in my time, and then some, to earn my way up. So don't think this is going to get easier anytime soon. Physics is difficult to begin with. And it may be hard, or impossible, to convince some of the faculty you belong here. But you're a smart guy. That's why we wanted you in our program. I believe in you. I believe you can put this behind you and make it through."

I can't remember whether I said thank you. Or if I just thought it. I know his words filled me up in places that had been running on empty for a long time.

63

TOOK A SHOWER and tried to grab some sleep before Jessica came home from the Regional Medical Center in San Jose, her first nursing job after completing school and passing her boards. But all I could do was lie there and think about how I would tell Jessica that I'd been leading a double life, that I had a dark secret I'd kept from her ever since JG and I headed down the crack hole together.

Dinner was always a protected space for us, so I waited until afterward to say what I needed to say. My somber mood was nothing new to her. She'd often say, half-joking, that I must be manic-depressive, since I was normally happy and high-energy, but during difficult times would descend into a deep funk.

After dinner, I stood at the sink washing the dishes in slow motion just to delay beginning the conversation. I worried that she'd be deeply disappointed in me, that I was going to break her heart, that she'd ridicule and demean me. We'd always had a contentious relationship. Rocky but committed. She just wanted a normal life with a man who worked a steady nine-to-five job. Someone who'd get along with her God-fearing family and live with her in a nice house in Natchez—or somewhere like Natchez. She knew I wasn't gonna be that, but she had me pegged as a quirky, relatively innocent nerd. I was about to shatter that image. To be honest, I was scared of her. Female anger always weighed heavy on me.

"Hey, I need to talk to you," I said finally.

"Okay, about what?" she responded, with a warm and beautiful smile she'd been wearing more since she'd gotten pregnant. Her upbeat mood made me feel even guiltier. I had to say it before I lost my nerve.

"I wasn't at the lab last night," I began.

Her face shifted quickly from light to dark. "Okay, so . . . where were you?" she almost whispered.

"I was in EPA. Doing drugs."

She looked a little relieved—which meant she must have thought it was another woman. "What kind of drugs?"

"The worst kind . . . rock cocaine."

"How long have you been doin' that?"

"Since Tougaloo. I stopped once we got married."

She didn't respond to that, so I continued.

"When you went back to Jackson in February, I started using again. I thought I would stop once you came back. Then I thought I'd stop after we found out you were pregnant. But I haven't been able to. It's been harder than I thought it would be to quit."

"So those nights when you said you were studying, were you using drugs then?"

"Not all the time. But when I did, it could go on for nights at a time."

"Sometimes I knew you weren't telling me the truth. I feared you were cheating on me," she said, and I could see her eyes tearing up. "Were you?"

I took her hands in mine and looked her in the eye. "I swear to you, I wasn't cheating on you. I was doing drugs. That's my big shameful secret. That's my only secret. And now I want to stop. I don't know how I'm gonna do it, because it calls to me every night. But I promise you I'm gonna stop somehow. And I'm not cheating on you."

She leaned in and embraced me, and I could feel her tears wetting my shirt. We held each other for a couple minutes, but I didn't cry. I didn't want to show her more of my weakness.

Jessica pulled back and put her hand on my cheek and said in her sweet but serious voice, "I'm with you. I know you can do whatever you set your heart on. We'll get through this together."

"Last night," I said, "a fourteen-year-old kid held a gun to the back of my head, and I thought I was dead. I was so afraid I'd never see our baby."

Right then, for the first time in the longest while, I felt the fear pass through me and then fall away like a spent rocket booster. I let the tears go too, crying like a little boy while Jessica rocked me in her arms.

64

THAT NIGHT, and over the next week, Jessica and I talked about how I was going to keep myself off the streets and away from drugs. The big question was, "Rehab or no rehab?" I was hesitant, because I didn't want rehab on my record somewhere, and then have it come out later in my career. I'd managed to avoid a criminal rap sheet, and I'd never been in debt. I didn't want a rehab record hanging over me. I knew that space research was tangled up with the federal government and the military, and there was no way I'd get security clearances if there was a record of me being addicted to crack.

Another reason I resisted rehab: I'd been trained since childhood to avoid going to doctors or hospitals except in cases of dire emergency. They were just too damned expensive. And rehab sounded even more expensive than normal healthcare. The good news was that for the first time in my civilian life I had health insurance. Stanford required it of all grad students, and because of my TA job, the school insured me.

I knew I needed outside intervention to put the demon rocks behind me. This *was* a dire emergency. I knew because the insidious voice of crack inside my head was so persistent it even drowned out my need to count. It was out of my control. It was controlling me.

Crack was a woman—don't ask me why, but she most assuredly

was—and every night when I slept, she took over my mind completely. She came to me in my dreams and displayed herself from every enticing angle, beckoning me into her grasp. She whispered in my ear promising me bliss, euphoria, and orgasmic release. She was irresistible. In my dreams, I always gave in happily and pulled her to my lips. Every. Single. Night. And it was so good.

In the morning, she retreated into the shadows. And as the day wore on, she started working me again. I bargained and rationalized. Access is my trigger, I told myself. As long as I stay out of the hood, I'll be fine. But every evening she would call to me, and every night, I would cross paths with her again in my dreams, and I'd wake up in the middle of the night sweating and afraid. I didn't know if I would ever be able to resist her. But maybe there was some way to kick without rehab.

I hesitated. Jessica did not. Not for nothing was she a trained nurse. She was no-nonsense and all business when it came to a physical addiction. She took charge and called someone at the Student Medical Center, who referred us to the mental health facility at Stanford Hospital. She made me an appointment to go in and speak with them.

I rode my bicycle to the Stanford Medical Center on the day of my appointment. The intake therapist was a skinny white guy with hair so curly, he could have styled a 'fro. His demeanor was super calm, and he spoke in hushed, soothing tones. He assured me there'd be no medical record that I'd attended drug rehab. When I told him I couldn't afford to be locked away for thirty days or whatever it would take, he said they had an eight-week outpatient program he thought would be appropriate for me.

Then he started asking me about my feelings. How did I feel about rehab? How did smoking crack make me feel? It was a strange question for me to hear. No one in my life had ever asked me, "How do you feel?"

It was as if I didn't understand the question because I reflexively replied, "I think . . ."

He listened patiently while I explained why I thought I went down the crack hole and why I thought it was a problem. Then he repeated, "Yes, but how does that make you feel?" That question would bounce back at me a hundred different times over the course of rehab. It was weeks before I could muster an authentic answer.

GROUP SESSIONS WERE HELD twice a week in the evening, which of course was when my cravings were the worst. There was one other Stanford student in the group, an Indian undergraduate woman who'd been partying and drinking nonstop since arriving at college. Eventually, she'd woken up in the hospital after nearly dying of alcohol poisoning. There were two Stanford employees in their thirties who were addicted to alcohol and pills. And there was another alcoholic in his early fifties who was the CEO of a small tech company in Silicon Valley. He was Irish, and when he found out I had Irish mixed in with my African and Creole blood, he took me under his wing. The only other crack user in the group was a normal-ass-looking white woman. That's the first thing I learned in rehab: addiction wasn't a Black thing. The scientist in me already knew that, but I'd never actually known any white addicts.

The second thing I learned was that I couldn't get sober on my own—any more than I could get through my physics coursework without a study group. Rehab was hard—harder even than quantum mechanics—because it meant diving into feelings I'd been pushing down for decades. I definitely needed teammates to show me the way and call me on my bullshit.

Each of us depended on everyone else in the group to show up at sessions and stick with the program. Partway through, one of the alcoholics relapsed. He was a skinny white guy who used to drink a fifth of liquor a night. When he relapsed and stopped coming to sessions, it shook us to our fucking cores. We were all on this journey of recovery together. When one of us fell off the cliff into oblivion, it reminded all of us how close we were to the abyss.

A few weeks in, we'd had hundreds of conversations about our addictions and how they made us feel. The Stanford therapists were no-nonsense types with no tolerance for anyone who didn't treat the process with the utmost seriousness. As professional and buttoned-up as they seemed, all the therapists had been alcoholics and addicts once themselves. So they understood what we were going through and all the ways we'd try to game ourselves and each other.

So they had my number. But for week after week, I felt like I was wasting my time. Despite all the talk, the crack-lady voice kept returning to my head, day and night. I wanted her as much as ever.

It was into week six when I finally learned how to block out that voice. Listening to everyone's stories, I realized they were all saying the same things:

When I use, bad things happen.

When I use, my dignity is compromised.

That's when it hit me: My demon lover promised I'd get high if I answered her call. But she lied. In truth, I'd get low.

Sure, I'd get high at first. But soon enough I'd be wandering through dark and dirty streets looking to get full. She'd bring me so low, I'd be crawling on the floor looking for dropped crumbs in the carpet like I'd seen my daddy stoop to doing, or selling my bed for a $50 fix like JG. That's not getting high. That's getting low.

So that became my mantra: "If I go back in, I won't get high—I'll get low."

65

W HILE I WAS in rehab that spring, I tried to figure out a strategy for successfully completing my undergraduate coursework. Being in group therapy several times a week convinced me that I couldn't excel at physics without a study group. But after being dissed by my classmates my first term, I was afraid of being burned again.

When Paul Estrada approached me after class one day and asked me if I was interested in studying with him, I was suspicious. He seemed weird, even by physics-student standards. Paul was a Latino who could pass for white and had science-nerd awkwardness written all over him. I politely declined his offer and dipped. But for some reason he persisted. The fourth time he invited me, I decided to fuck with him a bit. I stopped, looked him in the eye, and said, "Look dude, who the fuck are you exactly?"

Paul didn't blink. He came back at me with what sounded like authentic hood in his voice: "I'm the dude who going to save your ass, fool! So you with me, or what?"

I thought that shit was hilarious and busted out laughing. This quiet, weird-looking dude had a sense of humor, and some flavor.

"A'right, Paul." I nodded. "You got me." I lifted my right hand to dap him up. Evidently, he wasn't that hood. He put his hand out and hit me with a straight-up nerd-ass handshake.

When I showed up at Paul's dorm room at seven-thirty that evening, there was another guy I recognized from our class. A blond kid named Gavin Polhemus. Gavin was a white-bread nerd from the Northwest and one of the top physics undergraduates. He and Paul made an odd couple. The three of us made an even odder trio. And yet we fit together.

Like me, Paul was about four years older than most of the undergraduates. Sitting in physics and math classes with him, I didn't see our commonality. But he did—which is why he reached out to me in the first place. He was as much an outsider at Stanford as I was.

Paul's parents were solidly working-class immigrants from Mexico and Guatemala. Paul went to decent public schools in the Bay Area. But when his folks split up, he went off the rails. He fell in with a bad crowd in high school, started using drugs, and stopped going to classes. He claimed, proudly, to have skipped 107 classes his junior year. His father finally kicked him out of his house, and Paul had to finish high school through independent study.

For the next four years Paul became an outlaw, cultivating a big indoor grow and dealing weed by the pound in EPA. Eventually he began dealing coke and getting into scrapes with the law. He decided he needed to switch things up and applied to Foothill Community College. After bearing down and getting straight A's at Foothill for two years, Paul was able to transfer to Stanford. He'd only been on campus for a few months when we first met. Already, he was focused on an astrophysics track and was interning at NASA's Ames Research Center in nearby Mountain View.

Paul, Gavin, and I quickly became a power study group. Four nights every week we got together at Paul's place and worked on the problem sets. Not only was I learning physics at a new level, I discovered that when we worked together, I could stand toe-to-toe with Paul and Gavin.

We each had our own way of approaching physics. We called Estrada "Brute Force Paul," because he would solve problems without deploying the logical tricks that can save pages' worth of calculations. Paul loved to write all the calculations. Gavin was a double threat: he was mathematically as strong as anyone in the physics department, and he also had great physical insight. Though he was still an undergraduate, he thought like a seasoned physicist.

And then there was me. Working with Paul and Gavin, I discovered I could find my way through the maze of a problem intuitively, as if someone were whispering the path forward into my ear. Even though I wasn't as strong as Paul and Gavin in math, they were in awe of my problem-solving gifts. I was like the baby sidekick in the superhero posse who starts out not understanding his own powers yet, crashing into walls before he gets his bearings and learns to fly.

After bumping and dragging my ass through my first year, I soared through the spring semester of second year, finishing strong with three A's.

And then, my two amigos graduated and went off to PhD programs in physics—Paul to Cornell to work with Carl Sagan, and Gavin to University of Chicago to investigate string theory.

I was an academic orphan again.

66

MEANWHILE, JESSICA'S DUE DATE was fast approaching. We hadn't planned the pregnancy, but we were ecstatic about it. And now that I was through rehab and had my academic legs under me, I felt like I was ready to be a father.

After the disaster of Martel's birth and my breakup with Lisa, I was determined to do whatever I could to make this a happier beginning. We attended birthing classes two nights a week and looked for new ways to enjoy each other's company. Before confessing to Jessica and going to rehab, we'd been stuck in a cycle of stress. I was stressed by graduate school and my secret life on the street. I was often snappy or stuck inside my own head. To get back at me, she nagged and insulted me.

But then she got pregnant. I got through rehab. And we came together.

We did what young and hopeful couples do, and what I'd never had a chance to do with Lisa before Martel's birth: bought stuff for the baby's room and showed off her baby bump to friends. The folks at Berkeley's Center for Particle Astrophysics, friends from when I'd interned before Stanford, even threw her a baby shower.

HAVING WORKED AS A NURSE for years, Jessica wanted no part of a hospital birth. Given my experience with Martel, I agreed. Jessica wanted to give birth at home with a midwife. Since the Bay Area was ground zero for every facet of New Age living and healing, there were plenty of options. We selected a highly regarded woman not too far from campus.

But our baby had her own plan. The evening of May 29, Jessica went into labor.

When the midwife arrived at our apartment early the next morning, she checked Jessica's cervix and found that she had only dilated four centimeters. Throughout the day, the midwife came and left as Jessica's contractions increased in intensity. But her cervix wasn't dilating much. By the next evening, it became clear we'd have to go to the hospital.

After monitoring the baby's heartbeat for several minutes in the emergency room, the attending physician said, "The baby is in distress. We have to take her out now." A nurse quickly took my place at Jessica's side. They wheeled her into a surgical delivery room. I was given a gown and allowed to follow but directed to stay against the wall. I imagined how alone Jessica must have felt at that moment.

They cut through Jessica's abdomen, pulled our baby girl from her womb, and took her over to a side table and began working on her. I literally couldn't breathe until I heard our baby girl's cry. A moment later, a nurse placed little Kamilah in my arms, wrapped in her swaddle blanket. I looked down at her sweet little face and saw both Jessica and me. She blinked once and I saw that her eyes were grayish blue, just like my daddy and my two brothers, Byron and Fionne. I reached down and kissed her little forehead. My eyes watered with joy. I wanted to take her to Jessica and tell her, "Kamilah's here!" But Jessica still lay there unconscious and unmoving. I knew it was just the anesthesia, but I was anxious to see her awake and herself.

I sat there holding our baby girl and falling in love over and over as we waited for Jessica to awaken. Finally, she opened her eyes and I showed Jessica our little girl. She looked down at Kamilah and her tears began to flow. I sat beside her with my arm around her shoulders as she unwrapped the baby, lowered her gown, and placed Kamilah at her breast.

We were a family at last.

67

AFTER COMPLETING TWO YEARS of undergrad coursework in math and physics in strong fashion, I was finally prepared for graduate school. I'd made it onto Art Walker's research team and was working alongside Art and his grad students as we prepared for the next rocket launch. I had a happy home life with my wife and new baby. And I had kicked crack. For the first time, I had a life I could build on and a clear view of a future I could work toward.

But to succeed in the high-performance, high-stress environment of physics graduate school, I needed a physical outlet to help me blow off steam. I also needed an academic ally, now that Paul and Gavin had graduated.

Basketball turned out to be the fix on both fronts.

As soon as I arrived at Stanford, I'd discovered the basketball court right next to our apartment in Escondido Village. Every afternoon I knew I could find a four-on-four pickup game. Playing ball let me escape my head, have some fun, and just be in my body.

When I first started playing ball back in Piney Woods in my early teens, we didn't have a gym to play in or even a blacktop. So we built our own court. All we needed was a rim, with or without a net, a scrap of plywood to nail the rim to, and a sweetgum tree to sink into a flat spot of ground. After that it was just a matter of playing ball until our feet eventually hardened the ground and stripped it of all plant life.

I had zero skills back in my Mississippi playing days—just a lot of energy. I had a fast first step and I could jump, so I got a lot of rebounds. And I was always aggressive and competitive, which gave me opportunities to steal balls and make shots. I played plenty of ball in the navy and at Tougaloo, but I never stepped up my game till I got to Stanford.

After rehab, basketball became much more than just a way to break a sweat and run off some excess energy. It was an essential part of my recovery. It's how I quieted the taunting demon voices I couldn't seem to evict from my head. The voices that told me that I was a fraud. That I didn't belong in the same league with the ultra-educated set on campus. That I was just a temporary visitor from the hood. Shooting hoops cranked down the volume on that chorus.

I'd run myself ragged on the court every afternoon. By summer, I was in the best shape of my life. I could run down any player, go end to end for hours at a time, balling all the way. I could outlast anyone. What I lacked in talent, I made up for with athleticism, hustle, and a flair for trash talk that took all but the most strong-minded players out of their game. But it was all in play, and it was always fun. The only physics problem I had to solve on the court was finding the right trajectory for putting a ball through a hoop.

BASKETBALL BECAME MORE THAN just a physical and emotional outlet for me the day Daveed showed up at Escondido Village. He immediately stood out from the usual assortment of pickup players. Daveed was a chubby guy who you'd never imagine had game. But he had long arms, quick hands, a wickedly deceptive array of moves around the bucket, and a killer three-point shot. He spoke with a heavy Puerto Rican accent, peppered with Spanglish slang I didn't know—which made me think he'd grown up playing street ball.

We played our first game together one August afternoon before my graduate classes started up, and we exhausted ourselves testing and besting each other. I hit him with the kind of trash talk that would rattle most players. It rolled right off him. And when I moved in to foul him, he simply spun away and took the inside-under route to the basket for a quick two. We were having such a good time laughing and one-upping each other that when the game wound down to three-on-three and eventually one-on-one, we stayed and played until it was too dark to continue.

It was only afterward, when we were sitting sweaty and exhausted on the lawn courtside, that he told me he was starting grad school—in the physics department! When I heard he'd moved to Palo Alto straight from Puerto Rico, I asked him about racial divisions on the island. He said they weren't as bad as on the mainland. "But don't get me wrong," he said, "there are plenty of Puerto Ricans who think they're white and better than others."

I looked at his white-looking ass and asked, "What the hell do you think you are?"

"Hey man!" he protested. "I ain't white. I'm a Boricua!"

"You look pretty white to me," I said. "But I get it. You down with the brown."

"That right," he replied.

68

WHEN CLASSES STARTED, I couldn't help but notice that the other first-year grad students were on Daveed like white on rice. I didn't understand, at first, why they seemed to be hanging on his every word.

One day when we were sitting courtside, I asked him if he'd decided on an area of research yet. It turned out he'd already teamed up with Professor Robert Wagoner, a theoretical particle physicist and cosmologist whose book *Cosmic Horizons* had famously reconstructed the chemical makeup of the primordial universe, circa the Big Bang.

The fact that Wagoner had already accepted Daveed, a first-year newcomer, in his research group told me he was a heavyweight. The "third-floor theorists" as they were known, since they were housed in the third floor of Varian, were notorious as academic elitists. The rumor was that they weren't interested in working with American students, or minorities—only Russian, European, and Asian students need apply. Yet, Daveed had already penetrated their ivory tower.

As the semester continued, I'd only see Daveed on the basketball court, as he pretty much stopped going to classes. It was a habit he'd developed in college. Daveed grew up solidly middle-class in San Juan. Both his parents were chemists, but his father wanted to live in the neighborhood of his childhood, Catania, which was the roughest hood in the capital. That's where Daveed learned to play street ball. He didn't learn to speak English till he transferred to San Juan's American School in ninth grade.

During his freshman year at the University of Puerto Rico, Daveed took a physics class taught by Ronald Selsby, a Jewish physicist from the Bronx who decided to make his career in Puerto Rico because he liked drinking rum and fishing. He recognized Daveed as a super-smart kid with a talent for physics—but an aversion to attending classes. So he offered to tutor him privately. Over the next four years Professor Selsby taught Daveed everything he knew about physics, which went way beyond what was in the textbooks.

Tutoring and mentoring were habits Daveed brought with him to Stanford—along with an impressively deep understanding of physics. I guess he decided to pay his debt to Professor Selsby forward by sharing his love and knowledge of physics with anyone who asked. As soon as I started talking physics with Daveed, it was clear he was tuned into higher frequencies. His preferred classroom was Antonio's Nut House, a dive bar off El Camino Real. Using beer mugs, coasters, and saltshakers as props, Daveed could illustrate obscure theories of particle or quantum physics I'd struggled for hours to learn from a textbook.

SINCE MY EVENINGS were reserved for Jessica and Kamilah, Daveed and I fell into a daytime routine. I'd show up at his place at nine in the morning, and he'd make us Puerto Rican espresso on the stovetop using ground beans his grandmother shipped to him from San Juan. We'd spend the morning studying and debating physics, then head to the gym to lift weights. After a quick lunch together, I'd go off to work with my research group or go to class. Late afternoon, we'd meet up again for basketball and play till dinnertime.

Daveed and I both felt like outsiders. He was the only Latino in his class, and a long way from San Juan. He missed everything about Puerto Rico—the language, the food, the music, and his family. And he had the same problem as I did with all the privileged snobs at Stanford who thought they were smarter because they spoke what they considered "better" English, or because they had traveled around the world since they were kids. He liked to make fun of the squares and elitists as much as I did, even though his special physics smarts already placed him in the program's top tier, intellectually.

Maybe that's why I opened up to Daveed. When I told him about my misadventures in EPA, and about my hazardous romance with the rocks, he didn't judge me. Growing up in a tough neighborhood like Catania, he'd seen all sorts of bad behavior up close. He told me nothing I said could shock him. And I believed him. Art would check in with me about my rehab occasionally, just to make sure I was still on track. But he didn't want to be my coach or my confidant. So it felt good to be straight with Daveed.

The more time I spent with Daveed, the more at home I felt at Stanford—and the more confident I became with my grasp of physics. I trusted him enough to ask him any question, because I knew he was never going to look down on me. He recognized that I came at science from a different angle from the other students in our program. But he never made me feel like my language or my approach to problem-solving was any less smart than the Russian students in our program who'd been cherry-picked in kindergarten for a special track in science.

Daveed always told me, "There are a thousand ways to understand a physics problem and a thousand different paths to the right solution. You just have to find your best path." I loved that concept—that there were a multitude of pathways to a desired destination, that you could get lost in the woods and still find your way home if you had enough imagination and determination. I'd taken more than my share of wrong turns on what felt like a long and lonely journey. It was wonderful finally finding a fellow traveler.

69

NEEDED ALL THE SWAGGER I could muster to carve out a niche in Art's research group. For his grad students, it was all about taking ownership of a major problem in solar physics or a class of structures on the sun's surface. We all wanted to make our mark in some subspecialty that could launch our careers.

During the mid-'90s, Stanford physics professors won the Nobel Prize four years in a row. That's like the Chicago Bulls three-peating *twice* as NBA champions during that same decade. Totally dominant. And competitive as hell. That was the level of the game I was expected to play at.

Art's team was packed with students who had a running start in the race for a major physics prize. Half of them grew up as university brats on campuses where their dads taught physics and ran research labs. They got fed a daily diet of physics along with their Cheerios at the breakfast table. Tom Willis's dad, Bill Willis, was a particle physicist at the CERN accelerator in Geneva and winner of the Panofsky Prize. Max Allen's father and grandfather were renowned Canadian physicists. Craig DeForest had the most serious lineage of all: his father was a physics professor at UC San Diego, and his great-grandfather was Lee de Forest, known as the "father of radio" and the "grandfather of television" for inventing the Audion vacuum tube in 1906.

Everyone in Art's research group took teamwork seriously—

successful rocket launches depended on everyone nailing their assigned tasks. But that didn't mean all team members were equal. Seniority was the main determinant of rank. As the newest team member, I was at the bottom of the hierarchy, and senior grad students were at the top. I only started to feel accepted by my teammates when they began consulting me on their research ideas and recruiting me to help them with programming or coding problems.

As the low man on the totem pole, I was shocked when Art approached me to work with him directly on a project. The day he asked to meet with me in his office, I thought it was to check in on my progress in rehab. Which is where he started.

"So how's that going?" he asked me when we'd settled into our seats on either side of his monumentally cluttered desk. "Are you continuing to make progress with your healing?" Art always had a proper, gentlemanly way of talking.

I told him I was staying out of trouble, and I looked him straight in the eye when I said it. I didn't tell him that on any given day, the rocks might start calling to me. And I didn't tell him about my persistent dark vision from living the dirty life for so many years. Even in upscale Palo Alto I could see it happening all around me every day. People selling drugs and getting high. They weren't dealing and smoking crack on corners. But whatever the drug, whatever the hustle, it was the same dirt, and I couldn't stop seeing it wherever I looked. But what I told Art was, "I'm keeping my mind focused on MSSTA II," which was what we called our upcoming rocket launch, "and I'm staying home at night with Jessica and Kamilah."

"Good," he said. "I'm glad to hear you're keeping with the program. I've been tracking your work here in the lab, and I'm favorably impressed. I could use your help on a couple of publications I'm working on."

I was knocked out. He was offering me the two things we all wanted most: to work directly with Art, and to publish. Art only published in the highest-impact journals, so to co-author with him meant instant credibility and authority.

The fact that Art turned to me for assistance meant everything to me. Not only did he respect the quality of my work in the lab and with the group, he trusted me to show up and deliver—despite what I'd been up to a year earlier. I also think Art felt paternal toward me since I was the junior member of the team, and the kind of Black grad student he tried to convince his colleagues to accept for years. His own father was battling cancer at the time, and we all noticed the usually reserved Art was wearing his emotions on his sleeve for a change.

ART AND HIS ENGINEERING COLLEAGUES had solved one of the fundamental challenges in viewing the on-disk solar corona: the mirrors in telescopes can't reflect the extreme ultraviolet (EUV) light and soft x-rays that the corona emits. But astronomers learned that if they turned their mirrors ninety degrees, they could "skip" the light off of mirrors and acquire images, just like a rock can bounce off of water when thrown at a shallow angle. However, those images were of poor quality and yielded little meaningful data—unless you made the telescopes much, much larger, which would be very expensive. What Art and his team figured out was how to apply a coating to normal mirrors so they would reflect the EUV and x-ray light spectrum. This enabled telescopes to acquire high-quality images that revealed all sorts of previously washed-out details of coronal loops, rays, and plumes covering the sun's surface.

When Art presented his first solar images using his new multi-layer mirror technology, some of his colleagues celebrated his achievement, while others challenged the authenticity of his images. Art's findings were groundbreaking and, like every big leap forward in helioastronomy going back to Galileo, they aroused doubts among some of his peers. It was particularly hard to convince an international community of skeptical solar physicists of a breakthrough in their field if the pioneering scientist was a Black man who hadn't trained at a big-name school.

Art wanted to publish two papers to directly address the doubts of his skeptics. One paper would describe how the telescope mirrors and filters were designed, and the other would focus on the sun's radiation. He asked me to conduct a quantitative analysis of our data that would rebut the naysayers. I was thrilled by his invitation to collaborate. I'd have happily transcribed the stock-market page of the San Jose *Mercury News* if it meant publishing with Art.

After we'd been working for a few months on the two papers, Art said he'd be relying on me in the summer ahead to help prepare for the MSSTA II launch. He also talked to me about his father, and how after he got sick they were able to reconcile some long-standing conflicts. His father was terminal, and Art had moved him into his house so he could tend to him personally. Other times he shared news about his wife and daughter, and asked me about my family.

I didn't want to tell him about my daddy, who was still on the pipe. And I couldn't bear to tell him about Martel. But I loved showing him pictures of Kamilah and telling him how being a dad and taking care of a helpless infant had grown my heart by two sizes. I'd never experienced such a simple and loving relationship before. It went beyond the basic arithmetic of a parent giving and a child taking. It reminded me of Kepler's laws of planetary motion—how a planet's elliptic orbit around the sun is defined by their entwined gravitational fields. I didn't want to say it aloud to Art, but I was feeling the same kind of orbital bond with him that I'd lost with my daddy.

70

AT THE BEGINNING of the summer, Art found me in the lab and motioned me to follow him. "Come with me. I need to show you something."

He led me down the hallway to a small room with several large boxes sitting in the middle of the floor. Inside the boxes were dozens of pink bags filled with aluminum parts. Art pulled a giant loose-leaf notebook from a desk drawer and opened it on the desktop. Each page had a precise mechanical drawing of a part—beautiful renderings I immediately recognized from the times I'd interrupted Art in his office, bent over his drafting table.

"I need you to assemble the MSSTA's truss this summer," he told me. The truss was the skeleton that Art had designed to hold MSSTA's payload of nineteen telescopes securely in place while they hurtled through the upper atmosphere at seven miles per second. The freshly machined parts had just arrived directly from the machine shop. There were literally hundreds of them. Most were smaller than my hand. But there was no assembly diagram. Just hundreds of parts that I had to figure out how to fit together.

Not all the pieces were properly machined. So after comparing them to the drawings, I had to take some back to the machine shop and modify them myself. There were also no nuts or screws or bolts to fasten the parts together, so I had to measure and order all those too. And not just any nuts and bolts. For space research everything has to be done a particular way. If any bolts came loose during the launch, they would ricochet through the payload like a bullet as the rocket blasted through the stratosphere.

I also had to maintain rigorous cleanliness protocols. If any grease or oil from my hands or any other residue soiled the parts, these compounds would evaporate when the rocket entered the vacuum of space and condense on our mirrors, blocking out the EUV and x-rays we were trying to capture on film.

It meant a lot to me that Art was willing to trust me with such a critical component of the launch. MSSTA II's payload of telescopes and newly engineered mirrors was scheduled to launch in the fall. I understood that the success or failure of our mission could hinge on my performance, that my work on the truss was every bit as vital as that of the senior grad students who were programming the electronics boards or focusing the optics.

Art made it clear that he expected me to show up at the lab each morning at eight A.M. that summer. Most of the time in a research group we worked when we wanted, as long as we got the job done. So this was an adjustment for me. For the first couple of days I showed up at eight. Then, on the third day, after a night of walking figure eights around our apartment while rocking Kamilah back to sleep, I didn't show till noon—and Art chewed my ass out. If I'd been more mindful of Mr. Cross's dictum—"Discipline is the training that makes punishment unnecessary"—I could have spared myself that reprimand.

Art worked with me for about a week on the truss, at which point he told me, "You got this," and left me to sort it out on my own. Ever since my first research project, I enjoyed receiving an assignment and then going off on my own to complete it. But I'd been able to complete those assignments in a couple of weeks. This one took three months of long days working in a small room, phoning vendors to order parts, and going to the machine shop to modify incorrect parts or build missing parts on my own.

This was my biggest research test so far. Ever so slowly the truss started to take shape. It was an intricate jigsaw puzzle that took me directly inside Art's imagination. The truss was something he had painstakingly conceived of on paper, and I was bringing it to life in three dimensions.

As my work progressed over weeks and months, I began to sense how Art's mind worked and how the pieces would fit together. I lay in bed at night visualizing the half-built truss and seeing unattached pieces floating into place. I couldn't wait to get out of bed the next morning and return to my one-room lab and continue the construction.

Finally, every part, every screw, and every bolt was in place. Art came in and inspected each strut and fastener on the truss. It took him a full day of examination until he was satisfied that I'd executed his plan to a tee. He smiled at the truss, and then at me.

"Good work, James," was all he said.

But that was all I needed to hear. I'd passed the test.

71

IF YOU'RE TAKING PHOTOS of the sun using extreme ultraviolet light you need to launch your telescope and camera above the Earth's atmosphere. And to do that, you need a rocket with enough velocity to overcome the Earth's gravitational forces and make it into space. And if your payload is nineteen telescopes and cameras, you need a solid-fuel rocket that can generate enough thrust to accelerate to 20,000 miles per hour and clear the atmosphere. And if, in 1994, you wanted to launch a solar research "sounding rocket" with NASA funding, your best launchpad was the Department of Defense's largest open-air testing area: the White Sands Missile Range.

Near the end of World War II, the secretary of war carved the White Sands Proving Ground out of five thousand square miles of desert in southern New Mexico. On July 26, 1945, less than two weeks before Little Boy was dropped on Hiroshima, the Trinity Project team detonated the first uranium-based atom bomb in White Sands. And not long afterward, the War Department transported hundreds of captured German V-2 rockets and parts to White Sands for testing. Along with the V-2s, the United States also imported the V-2's Nazi architect, Wernher von Braun, and several of his top rocket scientists. Over the next decade, Von Braun and his team test-launched dozens of V-2s in White Sands and developed the first American ballistic missiles and rockets, including the prototype for the Saturn V that would power the 1969 moon launch.

IN LATE SEPTEMBER OF 1994, Art Walker and the rest of the team ferried our payload of telescopes and electronic boards down to White Sands. As the junior member of the group, I stayed behind at Stanford and ordered, machined, and Fed-Exed them any parts they needed as they assembled and tested the payload. Focusing and aligning nineteen telescopes with customized mirrors and micron-thin filters was complicated business, and by early October, we were weeks behind schedule.

In late October, two weeks before our new launch date, I drove down to join the team at White Sands. As I approached the missile range, I couldn't see anything but mile after mile of scrub desert. I had to stop and register at an entry gate guarded by a heavily armed sentry and show my birth certificate to prove my American citizenship. They gave me a photo ID for the range, then showed me a map of our designated Launch Complex 38, along with clear instructions about where I was and wasn't allowed to travel.

"Don't go down that road," the sentry told me, pointing to a road heading to a launch complex next to ours, "or we will shoot you." He gave me an "I'm not kidding" look to make sure I heard him.

When I entered the hangar inside our launch complex, the first thing I saw was the rocket itself—thirty feet of sleek titanium and aluminum—suspended horizontally above the workstations inside the bay. And I was pleased to see the truss I'd lovingly constructed packed with its payload of telescopes. The hangar had a high-ceilinged white interior with boomy acoustics, just like every sci-fi movie I'd ever seen with a rocket-ship hangar. In contrast to the casual atmosphere at our lab back on campus, everyone was intensely concentrating on his area of responsibility. Max Allen was tweaking the electronics boards. Charles Kankelborg was programming code in front of a computer terminal. Craig DeForest and Richard Hoover were focusing one of the telescopes on a U.S. Air Force resolution test pattern projected by a sun simulator.

With our new launch date closing in fast, everyone was on edge. This launch had been three years in the planning and execution. For a month now, the team had been focusing and refocusing the telescopes, and testing and retesting the payload.

Art had the most riding on the success or failure of the launch. EUV photography of the sun was his brainchild, and he'd raised the money from NASA. All his senior grad students were writing their PhD theses on different aspects of the launch and the solar data we were capturing. If you thought about all those telescopes with their precisely calibrated mirrors and filters being subjected to the g-force of a rocket launch fighting against the gravitational force of the Earth, you could go crazy. It was obvious that Art was carrying all the problematic scenarios in his head and all the stress on his shoulders. He was more tightly wound than usual. The rest of the team was uptight too, bickering and fussing with each other over every tiny task.

I was mostly there to observe my first rocket launch, while also trying to make myself useful by patching together whatever needed fixing and generally doing whatever I was told. I'd jump in and help whoever needed another pair of hands or eyes to focus a telescope or run a check on the code for a computer program.

The afternoon I arrived, we conducted the initial "shake test," which involved violently shaking the rocket and payload to simulate launch conditions to see what held together and what pieces shook loose. The days that followed were filled with an endless list of tests and tasks to perform.

As the launch date drew closer, we had to work more or less around the clock. Eventually, we moved a bunch of cots into the hangar and slept side by side in a tight row. With each passing night, we learned more than we wanted to know about what each person smelled like, and what kind of noises he made in his sleep. As the tension built and our nerves grew edgier, we more and more resembled a submarine crew that had been cruising at depth for weeks and was now impatient to engage the enemy.

72

THE DAY BEFORE the launch, we'd checked and rechecked all the systems. Dennis Martinez had finished reprogramming the ground-support equipment that would track the data flow from the payload in flight. There was nothing left to do but hope for good weather and good fortune the next day.

White Sands was administered by U.S. Army and Navy units, with NASA specialists on hand to help supervise telemetry and other technical areas. I was chatting up the navy engineer who was managing the rocket part of our launch. I told him I'd done a stint in the navy, and he told me about a pickup basketball game he played in at a gym near the entrance to the base. When I casually mentioned that I had game, he said, "C'mon and play, then. I'm heading over there now."

I sorely missed my daily b-ball game back on campus. With all the tension around the approaching launch, I was bouncing off the hangar walls. So I jumped into the jeep with the engineer and headed for the gym.

WHEN I GOT BACK an hour and a half later, still sweating from a competitive game of hoops, Art lit into me.

"Where the hell you been?"

"Uh . . ." I looked down at my feet. "The guys on the base needed an extra man for their b-ball game. . . ."

"The day before the launch, you're disappearing to the other side of the range? To play basketball? You make me wonder where your head is, 'cause it sure isn't where it needs to be, which is right here with your team doing final checks on the payload!"

I kept looking down, afraid to make eye contact with Art, or with anyone on the team who was listening to him chew me out.

"Here, make yourself useful," Art said, handing me a small metal plate and pointing toward the shop bench. "Drill these three half-centimeter holes in the spots where I've marked them. We're making washers to shim the telescopes. There's a drill press over there you can use."

I took the plate over to the shop bench and found the press, which I'd never used before or trained on. I didn't want to screw up, so I studied the press for a while and practiced drilling on some tin sheets. I couldn't seem to drill clean holes, maybe because I was still thinking about how Art had embarrassed me in front of everyone. But I sure as hell wasn't going to ask Art for help, and everyone else in the hangar was deep inside some task. I was still trying to figure out the damned drill press when I felt Art standing right behind me.

"What's taking you so long?"

"I've never used this kind of drill before . . ."

He looked at me the way my mama did when I'd taken apart a radio in the middle of the living room.

"Why didn't you ask for help then? You know we have a launch in eighteen hours, right?"

He shook his head like he didn't know what to do with me. Art was typically a gentle, patient guy. But that dude was gone. "I'll do it myself," he said, grabbing the drill from me and shouldering me aside. While Art checked the measurements on the holes and adjusted the torque on the bit in the drill, I just wanted to dissolve into the floor. One of my feet started tapping the way it did when I was stressed-out. I looked around for something to fix my eyes on and count. I needed something to do with my hands, which were twitching now. I spotted a pile of bubble wrap alongside the workbench. I pressed one of the bubbles between two fingers and kept on pressing as the air expanded against the plastic, and . . . POP!

Art startled and whirled around to face me, the drill still clenched in his hand. "Why the fuck would you do that? Right when I'm working with a piece of dangerous equipment?"

I stared down at my feet, but I could see with my peripheral vision that my teammates were looking over at us. Even the navy engineer had stopped what he was doing to watch. I thought I could hear someone snickering.

Art kept snapping on me and I just stood there stock-still. Finally, he said, "Just get out of my sight."

I turned my back on him and busied myself straightening a pile of plywood, just so it looked like I had something to do. I was filled with shame and embarrassment at being disrespected like that by the man whose approval I wanted most.

Art finished drilling the holes and held them up to his eye to make sure they were clean. I cleared my throat and said, "I'd like to talk to you in private."

Art glared at me, as if to say, *Are you serious? Now?* There wasn't any private place inside the hangar, so he marched outside into the blinding sun. I followed him a ways out into the scrub desert until he turned around to face me.

I was thinking I was about to stand up for myself and take Art to task for dressing me down in front of everyone. But Art had another idea. He started in with a tirade of Let-me-tell-you-all-the-ways-you're-fucking-up-and-driving-me-crazy. Why was I always waving my hands or scratching myself and doing weird things with my eyes? When was I going to get my act together? Why couldn't I just watch the other people on the team and do what they were doing?

After a while I stopped hearing the words he was saying and just squinted into the blinding sunlight bouncing off the white gypsum sand and listened for any other sound but his voice and tried to see anything but his face filled with reprimand and reproach. I held my hands tight behind my back and tried not to move, just letting the fierce white light wash over and around me.

But Art's voice broke through my white light shield: "James Plummer, you got a lo-o-o-ng way to go to prove yourself to me."

73

T HE NIGHT BEFORE the launch, Art assigned Dennis Martinez
and me to stay in the hangar and keep watch over the rocket and
payload, which was fully assembled and suspended horizontally
above the hangar deck. At dawn, the navy engineers would move it
into position on the nearby pad for our noon launch.

My job was to maintain the payload's vacuum overnight. I was
amazed that Art would still trust me with such an important assignment. If even a small amount of air leaked into the payload at launch,
it would cause a powerful pressure wave from one end of the payload
to the other, shredding filters and breaking telescope tubes. My job
was preventing a single-point failure that could ruin the entire mission. My anxiety level guaranteed I'd be wide-awake all night.

THE NEXT MORNING, we had time for one final check of the vacuum seals and the electronics system, including the "reset" sequence
Dennis had programmed into the ground support equipment. Finally, the navy guys towed the assembled rocket and payload out to
the launchpad and propped it upright. It looked like a sparkling silver spear in the gleaming sun.

Thirty minutes before launch, we all moved from the hangar to a
concrete blockhouse on the edge of the launchpad to watch the lift-
off. Even the roof of the blockhouse was concrete, molded in a

Mayan step-pyramid shape, so if the rocket misfired on launch—or if the first booster of the two-stage rocket fell straight down on us—we wouldn't be crispy crittered.

Three years of research and design and execution had all come down to a thirty-minute experiment. That's how long the payload took to launch through the atmosphere, collect its data from the sun, and descend back to Earth.

A rocket launch is all about achieving velocity. To overcome the Earth's gravitational field and ascend into the upper atmosphere, the gas exhaust exiting the bottom of the rocket has to generate an upward reaction force, or thrust. Our launch didn't look like what you saw on TV when a massive Atlas rocket propelled the Apollo spacecraft toward the moon. We had no big-ass cloud of gas slowly driving the big fuselage off the ground and into the sky. Our Nike-boosted Black Brant rocket just took off with a *whooooop!*

Our rocket shot straight up and up—until we saw the first booster turn off. There was a little explosion—*boof!*—and the Nike booster separated from the rocket that carried our payload and nose cone. Both parts continued drifting upward for a moment due to their inertia, and then—*swoosh!*—the Black Brant fired, and the rocket took off again and kept climbing while the Nike booster turned and fell back toward Earth.

A few minutes later, when the rocket became a small round fire-ball in the sky miles overhead, Dennis and Charles fixed their eyes on the instruments inside the blockhouse to make sure the payload was ascending on course and the electronics and cameras were switching on like they were supposed to.

I could tell from their worried expressions and quickly exchanged readouts that the electronics were failing to transmit data. "Time to reset!" Art shouted at Dennis and Charles. "Now!"

They punched in a command that fired a reset signal up to the payload. We all clenched our sphincters and held our breath.

"It worked!" said Charles. "We're getting data. . . ." It was too soon for high fives, but at least we could exhale.

About fifteen minutes into the launch, the instruments told us the payload was falling back toward Earth, right on schedule. The parachute deployed as designed, just after the nose cone reentered the atmosphere. But the wind had picked up enough to blow the payload downrange about sixty miles. Charles was tracking its descent from inside the blockhouse and had a fix on the projected landing coordinates in the desert.

Just then two large army helicopters touched down onto the launchpad. Charles, Dennis, and I scrambled aboard one of them while Art, Richard Hoover, and Craig climbed on board the other. After we strapped in, we put on headsets so we could talk over the noise of the engines and props. Flying over the missile range in pursuit of our payload, I realized that the main event at White Sands was military weapons testing. We research nerds were a sideshow.

When we flew over a mock town built out of plywood, I shouted to the pilot over the headsets, "What's that for?"

"That's classified," he replied.

Then we flew past a squadron of decoy tanks arrayed in formation and I waved down at them. "Classified." We could even see a giant U.S. Air Force resolution test target about a hundred yards in diameter painted onto the desert floor. Nearby was a series of deep bomb craters, which the pilot explained were from an antimissile battery firing on decoys launched from Hawaii, thousands of miles away. "Accurate to within twenty feet—give or take!" he shouted.

We finally spotted our payload up ahead, with its parachute lying alongside. The nose cone was still warm to the touch. We hustled out of the choppers, hauled the payload on board, and went aloft again.

As soon as we landed back at our launch complex, Richard and Craig grabbed the film rolls from the telescope cameras and retreated into the darkroom. It would take days to develop the film from all nineteen cameras, but we all waited nervously while they developed the first two rolls. When they emerged a half hour later with wide grins and two thumbs up, we could finally let out some whoops and cheers.

Art took us all out to dinner at the best restaurant in Las Cruces. It was a celebration with beer and tequila and a toast by Art saluting us all. Everyone was in relaxed good spirits after weeks of tension. Everyone but me. I couldn't shake the memory of Art dressing me down in front of everyone and telling me what a disappointment I was to him.

74

THE NEXT AFTERNOON, we crated and shipped the MSSTA II payload back to Stanford, packed up our gear, and headed home. I was driving solo for the seventeen-hour ride from White Sands to Palo Alto. Typically, I'd have driven straight through on I-10 past Los Angeles. But I wasn't in a hurry to get back home. Not until I sorted out what had gone down at White Sands. I wasn't ready to explain it all to Jessica, or myself.

I decided to return by two-lane highways, taking 210 North to Flagstaff, and then west on Route 40 through the Mojave Desert. Driving past Joshua Tree National Park after sunset, I decided to pull off the road and sleep outside in a sleeping bag I'd been using the past week during overnights in the hangar. The ground was hard, but there weren't any spiky plants or animals around, at least that I could see. I crawled into my bag and made a pillow out of my sweatshirt. It was a chilly night with a host of ice-bright stars poking through the moonless dark.

Lying out under the stars usually made me feel . . . not exactly godlike, but definitely part of something immense and magnificent. That night I felt tiny and useless. My dream of becoming an astrophysicist seemed further out of reach than ever before.

I hadn't shown up professionally for my team. They'd poured years of work into this launch, and at a key moment I'd lost focus. Worst of all, I'd disappointed Art. Mr. Cross's voice echoed in my mind: *Discipline is the training that makes punishment unnecessary.* It seemed to me that I was the one punishing *myself,* over and over. How long would this self-punishment be necessary?

I knew I could drive just two hours west and connect with my Crip cousins in South Central L.A. At least, the ones who weren't locked up. *You can score and be high by midnight,* a familiar voice called out to me. I let that voice play in my head for a while, to the point where I actually felt one of my hands unzipping my sleeping bag without my brain even telling it to. Before I could bolt for L.A., I zipped back up and lay there staring up at the stars, afraid to move until that demon voice had drifted off into the cold desert darkness.

But it wasn't that voice that spooked me. I'd grown accustomed to resisting its call, or at least ignoring it. I knew I was never going back down that hole. What scared me were the other voices rising up in my head, the ones asking the harsh, taunting questions. Would I always feel so alone and apart? Was there anywhere I truly belonged? Why was I always running away, and what was I running from?

I didn't find any answers in the night sky—just a numberless field of question marks. I closed my eyes, and when I opened them, the stars had washed away into dawn. It was time to get back on the road and head home.

75

A WEEK AFTER WE RETURNED from White Sands, the phone rang at my desk. It was Art. "James, come to my office, please."

I grabbed my lab book and a pencil and walked across the hall into his cluttered office. "Have a seat," he said, pointing to a chair piled with papers.

"I've heard from the Qualifying Exam Committee," he said. "You failed."

Qualifying exams, or "quals," are comprehensive tests graduate students take after a year or two of graduate coursework to demonstrate their overall knowledge of their subject. You have to pass your quals before you can begin work on your PhD thesis.

I'd taken the qualifying exams back in late September, just over a month before the rocket launch. I'd been so busy assembling the truss all summer, I didn't spend enough time studying for the quals. I'd done so well in my graduate courses, I assumed I knew all the material well enough to pass. I assumed wrong. A lot of grad students fail their quals the first time. Still, I was disappointed. And most of all, I didn't want Art to lose confidence in me, especially after our confrontation right before the MSSTA II rocket launch.

Art looked at me across his desk with his usual calm demeanor. His face and voice betrayed no emotion whatsoever. "The committee also had a message for you."

"Yes?" I said, hoping they'd sent me some sort of encouragement.

"The committee wants you to know that you have three options. I'll quote exactly what they said to me. Okay?"

"Okay," I replied.

"The first option, they said, is that perhaps graduate school is not for you, so you should drop out. The second option is that perhaps graduate school *is* for you—but just not at Stanford—so you should transfer. And the third option is you can stay and take the exam again next year. The committee strongly recommends that you take one of the first two options."

Art paused for a moment as I absorbed that gut punch. I'd made so much progress on all fronts at school, and this is what the faculty thought of me?

Art continued, "What are you going to do?"

I answered in a feeble voice, "Stay and take it again."

Art rose to his feet and exploded, "Great! Fuck 'em!" He pounded his desk with his fist. "Fuck all of 'em! I knew they were going to try something like this!" He paused, then looked me right in the eye and said, "I don't trust any of them."

Art was the whitest and most square Black dude I'd ever met. He had a framed photograph of himself shaking hands with Ronald Reagan hanging on his wall, for fuck's sake! If there was ever an establishment Black guy, he was it. And he'd just shouted aloud that *he* didn't trust any of them.

"There's something else they told me that is highly offensive, and blatantly against the rules. They said if you fail the exam a second time, there would be no oral and I won't be able to allow you to stay on, as is normally the advisor's prerogative."

I was taken aback on two fronts. First, that the faculty would deny Art discretion over the fate of his advisee. It was accepted practice that a student's advisor could request an oral exam if a student failed his quals twice, and the advisor had the final say over whether a student would be allowed to stay. I knew that Art had struggled to feel accepted and respected by some of his colleagues, but this was the first time I'd seen tension with his fellow faculty members ruffle Art's calm demeanor.

And second, I was personally hurt that the faculty would show such obvious disdain for my presence in the program. Of course, I'd heard from Professor Meyerhof and others that the younger professors who now chaired the faculty committees wanted the quals to have some "teeth." These were the same faculty members who resisted the department's program of having one or two diversity admits each year. Still, I didn't understand where this personal hostility was coming from. My experience in graduate school was nothing like my first two years with the undergrads. I was generally liked and admired by everyone. Why was I being treated this way?

Art took a deep breath and regained his composure. "You make sure you pass next year so I don't have to take these bastards on. I'm not having it."

"I'll do my best," I replied.

"Okay," Art said. "Please make an appointment to talk with Professor Wagoner, the Qualifying Exam Committee chairman. He'll deliver the same message to you. Just tell him what you told me."

I made the appointment with Wagoner. I then asked a couple of my friends who'd also failed the exam if they'd met with Wagoner, and what I should expect. They all had the same story: Wagoner asked them why they thought they'd failed and then told them to study hard so they could pass the following year.

That's not how my conversation with Wagoner went.

As I sat in a chair across from him, he presented me with the same three options that Art had. When I responded that I'd stay and take the exam again, he elaborated on the faculty's opinion of my prospects.

"Because you took two years of undergraduate classes, you've already been here a long time. By the time you take the exam again next year, you'll have been here four years. Usually when students leave because they're not good enough, they do so in their first or second year. The committee would like for you to complete all the requirements for your master's degree this year. That way, when you fail the quals next year, you won't leave here with nothing. Okay?"

"Okay," I responded numbly.

"Another thing is, you'll need a job next fall. At your level, it takes a full year to get a job. So the committee wants you to start applying for jobs now. That way when you fail the exam next year you'll have somewhere to go. Okay?"

"Okay."

I left Wagoner's office feeling defiant. *When* I fail next year? What the hell? Art was right. They couldn't be trusted. I'd worked hard and gotten good grades in my graduate courses. I was part of a team doing groundbreaking research. I'll show them! I thought. But another part of me felt the sting of rejection, like I'd been put in my place for even trying to gain acceptance to their club.

At least Art had my back. Art believed in me. And I was going to show him I was worthy of his support.

76

M Y RELATIONSHIP WITH ART evolved over the course of the winter and spring quarters. He started out the hard-ass task-master, insisting that I maintain regular work hours in my office, at SSRL, and in class. I treaded lightly in his presence and was sure to be fifteen minutes early to every appointment, just like in the navy.

My first assignment was completing the MSSTA II mission. Preparing, flying, and recovering our payload and data did not signal the end of the experiment. Postflight calibration measurements of all the MSSTA II optics needed to be completed at SSRL, where accelerator beamline time was competitively awarded. Art gave me responsibility for writing the proposal and serving as our group's spokesperson. I was elated when the proposal was accepted and assigned high priority, and SSRL awarded us twenty-one eight-hour shifts. That meant seven days of around-the-clock runtime. When I submitted the completed report to Art, his response was a mild "Thank you."

Gradually, hard-ass Art transitioned into wise-and-warm-mentor Art. He had me working closely with him on just about everything: drafting letters to colleagues; acting as his representative with NASA, SSRL, and Lockheed; and assisting him on his grant proposals. Meanwhile, I continued taking graduate courses and teaching Observational Astronomy.

Spending so much time together gave us lots of time to talk. He'd ask me about my progress in classes and if my family was doing well. Art was fond of using questions to instruct me. He'd frequently ask "Do you *know* that?" about whatever it was I claimed that I knew, "Or is that something you merely believe to be so?" It was his way of teaching me the level of evidence required to establish scientific facts.

ONE DAY ART ASKED ME, "Are you going to complete your master's this year, as Wagoner suggested?" Stanford's program was all about the doctorate. Some PhD students would pick up their master's just to say they had it. But most of the time it was considered a consolation prize for those who failed to earn a PhD.

"I am," I replied.

"Good," he said. "Are you inviting your family for the graduation ceremony?"

"No," I said. "I'm not going to attend the ceremony."

Art took off his glasses and sat back in his chair. "You're not attending the ceremony?"

"No."

"Why not?"

"Well, it's kind of disgraceful. It's a consolation prize," I said.

"Are you kidding me?" Art said.

I just shrugged.

"Has anyone else in your family earned a master's degree in physics?"

"No," I replied.

"Does anyone else in your family have a degree from a school like Stanford?"

"No," I replied again. I didn't tell him I was the first person in my immediate family to even graduate high school.

"What the hell is wrong with you? This is a big accomplishment. Don't let anyone make you feel bad about getting a master's degree. Invite your family. I'm excited to meet them."

Art was right about inviting my family. Mama, Bridgette, and Bridgette's two daughters all came out for graduation weekend, which is a big scene at Stanford. Bridgette and Mama took lots of pictures and Art made a big fuss about meeting them afterward. He was extremely gracious and charming, telling them how important I was to his research team and what a big contribution I was making. It made me blush, but it felt good to hear him praise me out loud in front of my family.

WHEN THE PHYSICS DEPARTMENT told me they'd be willing to hire a tutor to help me prepare for the quals, I requested Daveed.

The first night we met in the astrophysics library, Daveed proposed a strategy. "Look, bro, we need to focus you and get the biggest bang out of our time. I'm gonna shore you up on all the fundamentals. And the way I'll do that is through quantum. Twenty-five percent of the test is quantum, and those problems count for more."

"Whatever, Daveed. I trust you. Let's do this shit," I said.

"Understand, you're going to do all the work. Every day you have to read a chapter or two. Then that evening, you'll explain to me everything in that chapter in sequence as it's written. Then, you'll do every end-of-chapter problem in the book. After that I might give you some more problems."

"Well, stop talking about it and let's do it!" I responded.

"I'm serious, man," Daveed said. "We have to meet every evening. And I mean *every* evening. I'm not going to give you one day off."

"What the fuck, Daveed. Let's start tonight!"

Daveed stepped up in my face and spoke in the menacing growl of a drill sergeant. "I'm gonna break your ass down into pieces and rebuild you into the sharpest muthafucka in this muthafuckin' place."

DAVEED WAS SERIOUS about not giving me a night off. I thought I was going to lose my mind if I spent another night in that damned astrophysics library with him. But the system worked. Daveed was the best teacher I'd ever had. By mastering quantum, everything else was falling into place.

"You're ready, bro," Daveed said to me the week before the quals. "You're going to pass this thing."

"I sure as hell hope so," I said. "But man, I'm gonna study till the last second."

77

THE NEXT WEEK I sat for the two-day qualifying exam, along with a group of other grad students in the program. Each day we worked for four hours, took a one-hour lunch break, and then went back for another four hours. We waited for two weeks for the exams to be graded. Then one morning, word spread that grades were going to be posted in the physics building lobby that afternoon.

I joined the group of students already gathered in the lobby. After several minutes of waiting, I saw Daveed on the edge of the crowd, waving me outside. I found him next to the large wall of brass sculpture honoring Russell Varian, the physics building's namesake.

"Hey man," he said. "I have bad news. You failed."

I couldn't believe it. "How do you know?" I asked.

"I never told you," Daveed answered, "but I'm the student rep on the Quals Committee. I just left the meeting."

"Fuck!" Then it was true. My mind raced ahead to how I was going to handle my graduate school life falling off a cliff.

"But bro, listen: you didn't fail," Daveed continued.

"What?" I responded. "You just said I failed."

"You didn't fail, man. They're not kicking you out. You're gonna stay. I'll explain later."

Inside, there was a quiet frenzy as the graduate student coordinator, Marcia Keating, walked into the lobby and posted the exam results. There were two columns: one with a student ID number and the other with Pass, Fail, or Conditional Pass. And sure enough, I was listed as Fail.

As I walked away from the wall, Marcia called after me. "Please come into my office, James, so I can explain," she said.

WHEN I GOT TO Marcia's office, Daveed was there too. Qualifying Committee members are sworn to secrecy about their proceedings, and they both could have gotten in trouble for meeting with me privately. But as Daveed explained, "This was so fucked up, we have to talk to you about it."

Marcia put her hand on Daveed's arm to calm him. He was heated, as if they'd failed him too. "Here's what happened," he explained. "You needed to pass five of the eight parts. You passed four parts straight-out—including the three toughest: the two Quantum plus Stat Mech. You had a high enough score on the fifth section—but then Romani moved the goalposts so you'd fall outside the curve, and fail."

Now it made sense. Based on my interactions with Romani, who was chair of the Quals Committee, he seemed to be in the same camp as the department chair in his opposition to increasing diversity in the physics department.

"But the exams are anonymous," I said. "How could Romani single me out?"

"Bro, everybody knew who you were. Your handwriting, and more important, your distinctive logic chain gave you away. After four years, profs know your style."

"So what happened after Romani failed me?" I asked.

"Marcia brought the debate to a close," Daveed said. "She pointed out that you were grandfathered in under the existing rules. So if your advisor wants you to stay then you get to stay. That's when Romani said, 'Then, let's fail him, and let his advisor decide his fate.'"

"I don't care what anyone says," Marcia said to me. "You did not fail that exam. I was there in the room with the committee. They violated the rules to fail you. But Daveed defended you for passing the hardest sections and I worked out a compromise with Romani. So don't you worry about it. Art will have your back."

"What's he gonna do?" I asked. "They said that if I failed a second time, they were kicking me out."

"But you didn't fail! They just wanted to enter a fail into your record. You're staying. All you need is a letter from Art."

"Why are they acting like this?" I asked.

"Because they're elitist bitches is why!" Daveed shouted.

"This is done," Marcia said to me in a soothing voice. "Don't think about them anymore. There's nothing more they can do to you. From this point forward you just work with Art and move on. Mark my words, you're going to be one of the most successful people in your class. You have something they don't have. Everyone with a PhD can do physics. But you're a real person. Your personality, character, and imagination are going to take you places they can't go."

"Thank you, Marcia," I said. I left her office feeling uplifted. But now it was time to face Art.

WHEN I ARRIVED at Art's office, he greeted me with a smile: "Come on in!" I felt like I was about to ruin his day with my bad news, but he preempted me. "So I hear you passed the quals, but failed?" he said with a chuckle.

I raised one eyebrow. "You sure are taking it lightly," I said.

"They know you passed. Don't worry about this. It's nonsubstantive. I'll write a memo to them today saying I approve your staying. It's more than just a formality. It's their admission that they can't kick you out for nonperformance."

"Yeah, but it's still unfair," I said.

"Understand this. An organization of people is like a bell curve," said Art, citing a fundamental principle from probability theory. "In the center of the curve are the vast majority of people. They're indifferent. Apathetic. They're thinking of themselves. Then there is a small minority of people who will extend a helping hand to you. They will work with you, share resources with you. And there's another small minority that is going to be hostile to you. Don't let that small group of doubters derail you."

I wasn't convinced, and Art must have seen that on my face.

"Let me ask you a question," Art said. "How do you think the faculty treats me? Do you think they give me credit for my achievements?"

"Well, yeah," I said. "You are a full professor."

"And yet many of them still question my intelligence. They can never accept that a Black man is their intellectual equal or that you and I can make an original contribution."

I sat silently letting this statement of fact sink in.

"I love studying the universe and inventing new technologies," Art said. "How about you?"

"You know I do," I said.

Art continued, "I take pleasure in working with my colleagues—at least the ones who are not like these people. But the joy of my career is collaborating with my research students. I keep a special watch out for young scientists like you. I see how you think. And I see how hard you work when you apply yourself. What you have to realize is that in every group there will always be those who doubt you and try to make your life difficult. Screw 'em!"

I nodded.

"Now, let's get this data calibrated and figure out what the hell is happening on the surface of the sun."

78

As I began my PhD research, Art and I found ourselves in an international space race.

Just a year after our sounding-rocket experiment at White Sands, a consortium of European and American solar physicists launched the Solar and Heliospheric Observatory (SOHO) from Cape Canaveral. SOHO's two-year planned mission was to study the internal structures of the sun, its extensive outer atmosphere, and the origin of the solar wind—the stream of highly ionized gas that blows continuously outward through the solar system.

In the real world, science advances through a competitive international race to publish peer-reviewed findings before the other team does. That's how you make your reputation. Art and I were now in a flat-out sprint to publish our most significant findings from MSSTA II before we were eclipsed by SOHO's better-funded and better-equipped solar research expedition.

Satellites (like SOHO's platform) and sounding rockets (like MSSTA II used) are the tortoise and the hare of space research. Sounding rockets are good for deploying new technologies quickly and cheaply, and for seizing a snapshot of data in a few minutes. Rocket-launched satellites are good for obtaining higher-quality data continuously over a period of years.

Within eight months of its launch, the SOHO satellite would be in position in relation to the sun to start collecting and transmitting data. Two years after that, the international team of SOHO scientists would be analyzing and publishing their findings. Which meant we had less than three years to publish our own data to cap Art's pioneering career in solar physics—and hopefully launch my own.

By 1996, Craig, Charles, Max, and Ray had all earned their PhDs and moved on to postdocs or research posts at other universities and research centers. I was now Art's senior grad student and right-hand man. The week after the SOHO launch, he sat me down to strategize a publications plan.

"Luckily for us," Art said, "the low-hanging fruit is also our juiciest data: our high-resolution images and broad thermal coverage. We can be the first to determine whether plumes are the sources of high-speed solar wind emanating from coronal holes. I've worked up the underlying equations, but I need you to recalibrate the image data and make the spatial and spectral measurements that go into them."

"Yes, sir!"

I was pumped, because this research had profound real-world implications. Even nonscientists understood the significance of decoding the source of solar winds, which controlled space weather in our solar system. Satellites had become the linchpin of communications, meteorology, and defense intelligence-gathering. Solar storms—giant explosions on the surface of the sun—had emerged as the most dangerous naturally occurring event affecting our space technologies. In addition to disabling satellites and disrupting communications, solar storms create massive magnetic fields that could shut down power grids and disable aircraft flight control.

So the race to decipher the source and character of sun storms was also a race to protect planet Earth and its fragile electronic infrastructure. A real-life superhero adventure!

AT THIS POINT Art and I were resonating on the same wavelength. I already knew that Art's directive to "recalibrate the image data and make the spatial and spectral measurements" wasn't just a programming assignment. To have an unassailable list of citations for our articles, I had to become current with the literature. That meant reading every journal article ever written about solar plumes. Which meant burying myself in the journals section of the physics library and combing through current and back issues of *Solar Physics, The Astrophysical Journal, Astronomy & Astrophysics,* and *Monthly Notices of the Royal Astronomical Society.* Every volume going back decades and some into the previous century.

I checked out journals by the armload and took them to our office and lab complex to photocopy. I filled drawers and covered desks with photocopies of journal papers. My office became almost as cluttered as Art's.

But most of that material was now stored inside my head. If Art referenced a particular phenomenon in solar astrophysics, plasma physics, or quantum mechanics related to our papers, I could usually cite the latest publications by memory. As my expertise and our relationship deepened, Art treated me more like a colleague than a student. It was strange at first, but over time it started to feel normal. I'd become Art's trusted second brain.

79

Art wasn't one of those professors who socialized with his students or invited them over to his house for barbecues. He was a private guy, reserved and almost formal. Which is why it meant so much to me when Art started taking me out for beers after work. When a distinguished colleague was visiting campus and Art took him out to dinner, he brought me along.

Art wasn't just mentoring me as a scientist. He was connecting me to a larger network outside of Stanford. Sometimes it even felt like he wasn't just showing me the ropes—he was showing me off. He started inviting me to gatherings of Sigma Pi Phi, the oldest Black professional fraternity in the country, with membership reserved for highly accomplished individuals in the arts, law, medicine, and business. Art was one of the few scientists among the invitation-only roster of distinguished public figures such as Arthur Ashe, Vernon Jordan, and Congressman John Lewis. Martin Luther King had been a member. For those kinds of dinners, I'd make sure my one all-purpose weddings-and-funerals-and-graduations suit was clean and pressed. I even spit-polished my shoes, the way they taught me in the navy.

After an evening out together with a visiting lecturer from Princeton, Art gave me a lift home. Looking at him in profile, I could see his usually pudgy face was lean and drawn. His suit jacket was hanging off him. When I asked him if he'd gone on some sort of diet, he laughed at that idea. "No, I've just been having some stomach issues. I must be getting old, because I can't digest my food the way I used to."

Art was still in his early sixties, but I noticed he wasn't keeping up his usual tireless pace. He was typically at it day and night on whatever project he was pushing forward. His wife, Victoria, once told me she figured out what she'd signed on for during their honeymoon cruise to Bermuda. The second night out, she woke at three A.M. to find him crouched over a stack of papers. He worked every night of their honeymoon after she went to sleep.

I wanted to ask Victoria what was going on with Art's health. But they were both private people and I never saw her separate from Art. Then one day I ran into Victoria at the Safeway and decided to broach the subject. I told her I'd noticed Art wasn't his usual high-energy self. Was there something wrong?

She hesitated, then looked around to see if anyone was near us before she spoke. "Art's often told me he thinks of you as the son he never had. Which is why I'm telling you this, for your ears only. He's worried that if word gets out, the other faculty will start angling for his team's office and lab space. And you know how private he is." She must have seen the concern on my face, because she took my hand when she said, "Last month, Art was diagnosed with stage four pancreatic cancer."

My first thought wasn't a thought at all—it was a feeling of suddenly losing something important, as if a vital organ had just disappeared from my chest, and all that was left was a hole with intense feelings flooding through it. I was going to be Art-less, orphaned of my mentor. Art had been telling me, "You've got to build a life your daughter will be proud of." But I wanted to make *Art* proud as I set forth as a scientist and made my mark on the world. I felt the enormity of everything Art had learned and discovered and transmitted to me. How could all that mind and heart simply turn to dust?

Now I was racing against two ticking clocks: SOHO's publication schedule and Art's cancer. I didn't know what his prognosis was, but I knew we weren't just working to submit his latest round of journal articles. This was important to Art Walker's legacy—and the legacy of Black physicists.

In a moment of candor, Art had confided to me that despite all his accomplishments in solar physics, he was constantly fighting racism's legacy of questioning the intelligence of Blacks. In the world of physics, Art explained to me, some still believed that while Black scientists might be able to build ingenious gadgets, they weren't intellectually or mathematically gifted enough to make insights into the workings of nature—either in pure physics theory or in the analyses of data and observations.

Art's career had started off like gangbusters in both engineering and theoretical physics. But even though he was credited with developing the novel technology that enabled routine full-disk observations of the sun's corona—which was based on breakthrough physics—the elitists pointed out that Art had not published any astrophysics papers since his work with Sally Ride almost twenty years earlier. Among his many recent journal articles, only one was a pure "science" paper. The rest were deemed by the doubters as merely technical engineering publications.

Art badly wanted to publish findings from the MSSTA II data that explained the physics of how things work on the sun's surface. His goal was to advance the field's scientific understanding of the solar wind, the heating and creation of the sun's atmosphere, and the mechanisms of energy transportation through the atmosphere.

My mission was to deliver on this last chapter of Art's legacy while he was still alive.

80

HADN'T SEEN or spoken to my daddy in several years. I'd been afraid to visit him, though I'd heard from my half-brother Fionne that he'd been doing better. When Fionne invited me to a New Year's Eve party in New Orleans that year, I decided it was time to check in with Daddy.

He was living on his own in New Orleans East, down by the water. When I came to the door, he didn't invite me inside, so we sat on the front steps. I couldn't see much of him in the half-dark, but he seemed older and quieter than last time I'd visited. I told him I was doing well at school, and I invited him to come out for my PhD graduation that May.

"I'd like to do dat, sho' enough," he said. "A real professor, now. Dat right?"

"Almost," I said.

"Your Mama tells me you went to rehab. Dat right?"

"Yeah, that's right. I had to get outta that street life."

"Good for you, son. I done the same. Went and got straightened out. I'm a better man for it too. Dem rocks will ruin a man's life."

"I'm glad to hear you're doing well, Daddy. I'm proud of you for that."

We shot the shit for a while, and then it was time for me to go meet up with Fionne. When Daddy rose to say goodbye, his voice dropped to a whisper.

"I'm a li'l short this month. After Christmas and everything. I gotta make rent tomorrow, and . . . well, I'm short. Can you spot me somethin'?"

"Of course, Daddy." I had $130 in my pocket. I gave him all of it. Then I took off for the party.

It was an old-fashioned house party at a cousin's place. Folks played cards and danced, drank rum and Cokes and smoked blunts. I knew a lot of the people there. Most of them razzed me affectionately for "turning Cali" on them.

Sometime after midnight I caught sight of Daddy across the room. He was holding a beer in his hand and rocking back and forth, mumbling to no one in particular. There was no mistaking it—he was tweaked out. He'd lied to me about going straight.

I got up to go, but someone started talking to me about his brother-in-law who'd moved to San Francisco, and did I know him? Just then, Daddy looked across at me and our eyes locked. He saw me staring at him, and a look came over his face I'd never seen before. He was ashamed.

I beat it out of there. I couldn't bear to see him like that.

81

ART TAUGHT ME how to behave like a full man: How to be a professional scientist who showed up on time and did his job. How to be a gentleman who didn't lose his cool and could command the respect of his peers and his students. How to engage with people without triggering their tribal hierarchies. How to control my mouth and my speech, to know when to speak out and when to remain silent.

Most important, Art taught me the difference between what I believe to be true and what I know, as a scientist, to be true. That's the final measure of a researcher: to be able to challenge one's own beliefs and prejudices; to look at the facts with a clear eye and an open mind; to never fill in the blanks with one's own expectations but to let the evidence tell the story.

Having a role model for both manhood and a life in science was huge for me. But it wasn't the same as becoming a man, or a scientist. To do that, I still had to slay the demons who'd been chasing me since childhood. I had to unlearn my ingrained survival reflex to go tough and self-protective whenever I felt vulnerable or worthless. I wasn't going to be able to stop pushing the self-destruct and eject buttons until I repaired the parts of me that still felt broken. And I wasn't going to be free of my dirty vision of the world around me until I reclaimed my innocence and retrieved my heritage.

Back when I was at Tougaloo, I'd read the basic canon of Afrocentric literature. At Stanford, I became determined to discover how my people and I ended up at the bottom of America's identity hierarchy. I wanted to learn the history, religions, and cultures of Black people from Africa and beyond. Stanford's Green Library became my exploration ground.

Searching the sub-basement stacks felt like spelunking in a cave filled with treasure. There was row after dark row of books, and at the end of each row was a light switch on a timer. Once I found the book I was looking for, I'd sit on the floor beneath the cone of light and read until the light switched off. Then I'd find another book in another stack and read that for twenty or thirty minutes till that light switched off too. By closing time, I'd be sitting beside a pile of books deep inside some piece of submerged African culture.

When I decided to change my name, it had nothing to do with rejecting the religion I was raised in, or my daddy, or my past. I had always been proud of being James Plummer Jr. from New Orleans East and Piney Woods. For me, changing my name was about staking a claim to my identity. My ancestors were forced to change their names when they came to America as enslaved people, like Kunta Kinte did. They had no choice. I decided to change my name as a matter of choice and self-determination.

As I approached my thesis defense, I was ready to start a new phase of my life. I didn't feel reborn exactly, but I felt like I'd grown from a boy into a man. For three decades, I'd been getting by on my survival instincts, like someone driving through the darkness with only dim headlights to guide him. It had taken me a long time to overcome my childhood. I was almost thirty years old before I learned to value myself enough to stop seeing others as a threat, and to stop posing a threat to myself—before the boy born James Edward Plummer Jr. was ready to become a man. From that point forward, I claimed my future as my own.

If I could one day make a significant contribution to science, I wanted people to know, just from hearing my name, that I was a Black man, descended from Africa.

I wanted my first name to express who I wished to become: in cultures from North Africa to east of India, "Hakeem" means "wise." I wanted my middle name to express who I felt myself to be: "Muata" is Swahili for "he seeks the truth." I wanted my last name to be West African, where my African ancestors are from, and to have a noble meaning: "Oluseyi" is Yoruba for "God has done this." I wasn't bowing down to any particular god. But after everything I'd been through, and had put myself through, I figured it was time to start treating my life as something sacred.

As soon as I changed my name, I discovered people took it personally.

Everyone at Stanford congratulated me on my new name. They thought it was cool and fit me much better than James Plummer. They immediately started calling me Hakeem and most of them took the trouble to learn how to pronounce Oluseyi.

When I went back home to Mississippi over Thanksgiving, the response I got from a lot of folks was, "You still James Plummer to me."

Art's response was pure Art. "Okay, Hakeem Oluseyi . . . let's get back to work!"

82

ALL MY LIFE I'd been told that to be accepted by whites as an equal, a Black man had to prove himself twice as good. That was cutting it way too close, if you asked me.

As my thesis neared completion, I had authored thirteen published papers and was first or second author on eight of them. I had confidence in every element of the analyses and findings, which included the discovery and description of an entire new class of solar atmospheric structures. I was ready to present and defend my thesis—and earn my PhD.

I needed to recruit five professors for my dissertation committee: four from physics and applied physics, and one from outside the department. I was clear about who I'd ask to join my committee. First was Art, who as my mentor would be the committee chairman. Next was Phil Scherrer, Stanford's other experimental solar physicist, who built the Michelson Doppler imaging device on the SOHO satellite that revolutionized the measurement of solar magnetic fields. For the non-physicist on my committee, I asked Reginald Thomas, a Black professor in the engineering department. Finally, I wanted to add the two toughest professors in the physics department, including the professor no graduate student in his right mind would select: Bob Laughlin.

The Nobel Prize winner in physics the previous year, Laughlin was a dominant theoretical physicist whose areas of expertise ranged from cosmology to plasma and nuclear physics to nuclear-pumped x-ray lasers. I never wanted anyone to question whether I earned my PhD on its merits. So I asked Laughlin to join the committee to evaluate my PhD thesis, and he accepted.

Most folks regarded Laughlin as an intellectual bully, and his large and solidly built frame made him a physically intimidating presence. He was notorious for standing up in a packed auditorium during a physics colloquium and loudly challenging the distinguished visiting speaker. He was also known for consistently authoring the toughest quals problems. The first time I took the exam, Laughlin asked us to design an antimissile system that used frozen turkeys as projectiles.

I sensed that Laughlin was an outsider nerd, like me. He could get up in your face if he disagreed with you, but he seemed to me less like a bully than a guy playing defense against the doubters and haters. He'd been a star physicist and a nonconformist right out of grad school, which made him a target of resentment from his academic colleagues. I'd taken his undergraduate Statistical Mechanics and graduate Solid-State Physics courses, earning an A in Solid-State. Laughlin had always treated me with respect, even when he called me out for making mistakes. So I wasn't afraid to have him judge my thesis.

For my last physics faculty advisor, I turned to the professor who offered me the humiliating bail-out options after I failed the qualifying exam: Dr. Robert Wagoner. I wanted more than anything to prove to him how wrong he and the faculty had been about me.

ON THE DAY of my thesis defense, I entered the lecture hall with seven manila folders of transparencies—one for each chapter of my dissertation. The defense consisted of a public one-hour presentation of my dissertation research and findings—followed by a one-hour grilling by the committee members in a private session.

I began the open session with an overview of my research into the sun's "transition region" and presented each of my three published science papers on the topic that I had co-authored with Art. Midway through my presentation, Professor Laughlin stood up in the front row of the lecture hall and loudly challenged my assertion that an image of the solar corona superimposed on the magnetic field showed a spatial correlation. We went back and forth a few times before I finally said, "Look, Bob, you can see whatever the hell you want in it. But my quantitative analyses consistently show a strong correlation." Having tested my confidence in my findings, Professor Laughlin conceded the point and sat down.

After I completed the public portion of the talk and answered questions from the audience of physics students, the graduate student coordinator directed everyone but my thesis faculty to leave the room.

Now I was facing off against the committee members. They sat in a cluster a few rows away from me, speaking softly among themselves, chuckling occasionally but mostly bantering back and forth in hushed voices. I couldn't make out a word they said. But for a full fifteen minutes they ignored me as I stood in place. It required all my navy boot-camp and fraternity-pledge experience to remain internally centered and focused on the job at hand—which was to stand impassively and pretend not to be sweating through my shirt.

Eventually, the questions came—slowly at first, then rapid-fire from one member after another. I sensed they were testing my composure and my ability to stand behind my findings as much as my command of the subject. I stood my ground and responded to all their queries with the self-confidence that came from realizing that I was as expert as anyone in the room on this topic. I found myself flashing back to the state science fair and the feeling of power that came from batting back questions from the judges, gaining strength with each successive volley.

When the hour-long inquisition finally concluded, I was exhausted and exhilarated. Art said, "Go wait for me in my office." Then he turned back to the other committee members, and they resumed their collegial banter.

I paced nervously around Art's office, scanning the books and journals on his shelves but unable to focus my eyes, much less my mind. Eventually, the door opened and Art walked in.

He reached his hand toward me. "Congratulations, Doctor." Art's handshake, and the hug that followed, were all the affirmation I could ask for.

An unexpected encounter followed a few minutes later, when Professor Wagoner buttonholed me in the hallway outside Art's office.

"As a rule," Professor Wagoner said, a stern expression on his face, "I find that the length of a PhD thesis is inversely related to its quality." That made me squirm, as my physics thesis was the longest in the department's recent history. "However," he said, finally cracking a smile and extending his hand, "yours proved a notable exception to that rule. Excellent work, Dr. Oluseyi."

A YEAR AFTER MY GRADUATION from Stanford, Art and I co-hosted the 2001 Annual Meeting of the National Society of Black Physicists at Stanford. Art had lost a lot more weight and looked a bit like a stick figure inside his suit, but he managed to hold forth in fine form. Working alongside him to organize the meeting, and watching him preside proudly and forcefully over this congregation of his Black peers, felt like his final gift to me, and his final lesson in manhood. The highlight of the annual meeting for me was when Art presented the last of the thirteen publications we wrote together.

Art died a month after that conference, in April of 2001. I was a pallbearer and his only student and non–family member invited to his small family funeral. The public memorial service held for Art at Stanford's huge Memorial Church was standing room only. I shouldn't have been surprised by the accolades and awards that followed his death, but they reminded me of what a towering figure he'd become in our field.

In his last year, Art helped lay the groundwork for my postgraduate career. His wife, Victoria, was the one who encouraged me to take my first job at Applied Materials in Silicon Valley. But it was Art who steered me toward my next home at Lawrence Berkeley National Laboratory, where I joined the Supernova Cosmology Project developing detectors that would be selected for the Dark Energy Camera and the Vera C. Rubin Observatory.

Later that same year, my daddy suffered a severe stroke. He and I hadn't spoken since our last encounter on New Year's Eve two years earlier. When I went to visit him, I found a broken old man wearing a diaper, with a ring of white hair on his head. But he still had his personality and his sense of humor. And he was sober for the first time in a long while.

He asked me to take him to a juke joint he liked out on Highway 11, outside of Laurel. We drank RC Colas and Daddy talked to me about country things. Farming and working the land and all the things he knew more about than anyone I'd ever met. I realized how much of that Mississippi-grown know-how would disappear with him when he passed on in a few years.

Then a song he loved came on the jukebox, and Daddy sang for me in that soft, old-timey Mississippi blues style he had, and that I hadn't heard out of him since I was ten and we were walking together through the sugarcane fields on Kelly Hill.

That year, when I was ten, was my first sugar harvest. All summer the sugarcane had been off-limits to us kids. Cane was a cash crop. But like most poor kids who didn't have very sugary lives, I had fantasized about getting my hands on some of that cane to chew on. As the harvest approached, I planned to scavenge in the fields after Daddy and my cousins had cut down the stalks and shipped them off to market.

When I came home from Quitman Elementary the day of the harvest, I saw all the leftover stalks were laid out at the end of the field—and they were shoveling dirt on top of them! I was in shock by the time they finished up burying the stalks. Daddy tried to comfort me by explaining that in a few months there would be new sprouts at the joints of those cane stalks. Burying them at harvest time was how you planted the next year's crop. You couldn't grow next year's sugar unless you buried some of this year's harvest under the field.

Losing the two towering men in my life left me feeling alone and yet strangely freed. They had both taught me, in different ways, that self-respect had very little to do with self-confidence and everything to do with self-mastery and self-acceptance. They helped me start down that road. I had to finish the journey on my own.

I would forever embody attributes of both men, though they were now consigned to the realm of my ancestors. They would both speak to me, as ancestors do from time to time. But their presence would gradually grow fainter, like a binary star retreating from the planet it had spawned.

It was time to forge my own path into the future.

Epilogue

EMBRACING MY FUTURE as a research physicist meant putting the street in my rearview mirror. But I didn't want to leave my community behind. I knew I wanted to help the next generation of outsider scientists avoid the self-destructive cycle of fight-or-flight I'd been caught up in.

Not long after I left rehab, I began tutoring Black and Latino high school students at Carlmont High School in nearby Belmont. It was the same school where the movie *Dangerous Minds* was set, with Michelle Pfeiffer playing the former marine who went there to teach. I taught math to academically ambitious teenagers who had all shown up for after-school tutoring of their own accord. These were kids who realized that their education was falling short, and that honing their academic skills was their path to a better life.

They'd never had a teacher like me—which is to say, someone who looked and talked like them. I gave them a different perspective on matters of math, and something more: the rigor, self-confidence, and self-respect they needed to master challenging material. And I offered them a path toward their dreams that they could believe in and actually achieve. I'd found my personal power in science, and eventually I found my tribe among my fellow researchers—but not until I stopped embracing the negative stereotypes the culture kept projecting on me. I taught my Carlmont students to seize on their strengths instead and develop them to the max.

Sometimes personal strengths aren't obvious. They might even get you tagged as weird. Growing up, I compulsively counted the objects in my environment—partly to soothe my anxieties and partly to unlock the mysteries inside things by enumerating them. This habit earned me nothing but taunts and bullying as a child. I did my best to ignore the haters. Whenever I looked up at a moonless night sky, I wondered how I might one day count the stars. Years later, when I became a research cosmologist, I made one of my biggest contributions by developing detectors that can calculate the multitude of visible and invisible objects in our universe.

In 2002, shortly after earning my PhD, I began traveling to Africa to help educate that continent's next generation of astroscientists in Swaziland, Zambia, Tanzania, and Kenya. A few years later, when I'd become an astrophysics professor at the Florida Institute of Technology on Florida's Space Coast, the Kellogg Foundation gave my colleagues and me a five-year grant to create a mentoring program for Black astronomy students in South Africa.

Because of the unique viewing conditions in that part of the southern hemisphere, South Africa had always attracted worldclass astronomers. Historically, they'd mostly been British, going back to the early nineteenth century when Sir John Herschel traveled to the Cape of Good Hope to view the return of Halley's Comet and chart the stars and nebulae in the southern sky.

After apartheid ended in 1994, and many white astronomers exited the country, the new government recruited the top students educated in apartheid's separate and unequal schools—known as "Historically Disadvantaged Institutions"—to attend the elite National Astrophysics and Space Sciences Programme at the University of Cape Town. But almost none of these "historically disadvantaged" students could pass the honors exam.

At the time, I was the only astrophysicist in America who had graduated from an HBCU and had gone on to win respect as a researcher in the field. So it was no accident that the Kellogg Foundation saw me as a role model for aspiring Black astronomers halfway around the world.

At first, my South African students didn't identify with me at all. To them, I was an affluent American astrophysicist and professor. They had grown up in poor Black townships and villages. I had to show them I understood the two worlds they were struggling to straddle. Back home, they were heroes—the brilliant kids who'd gone off to Cape Town to become astronomers. But here at the university, they felt like under-educated, second-class students. I shared with them my own struggles to overcome those feelings of inferiority, win the acceptance of my academic peers, and master the rigors of high-level science.

I knew that if these students were going to compete with the best at the University of Cape Town, and later in the international scientific community, I had to set a high bar. So instead of drilling them on the material they'd be tested on, I decided to jump way ahead and teach them cosmology and quantum field theory—subjects considered the apex of physics. Once they conquered the apex, they'd know they could learn anything—and they'd believe in their futures as scientists.

And that's the way it went. Not only did every one of my students pass their honors exams, they all passed in the top 20 percent. The proportion of Black astronomy PhD students in South Africa spiked way up that year, and it's continued to climb ever since. There's no limit as long as kids keep dreaming. Our universe of an estimated hundred billion trillion stars is vast—but it's still finite, not infinite. The closest thing to infinity I've ever observed is hope. The infinity of hope is what I saw in the faces of my South African students.

Soon after we launched our mentoring program, South Africa entered and won the international competition to host the most powerful radio telescope cluster in the world, the Square Kilometer

Array (SKA). If you look at the photo of the South African SKA team, you'll see four of my African students smiling proudly in the front row.

I'm not in that photo, but believe me, I'm standing tall and proud right next to them.

Acknowledgments

WE'D LIKE TO SALUTE the constellation of talented individuals who piloted this book to publication:

Our agent, Howard Yoon of the Ross/Yoon Agency, introduced us to each other and championed this project from day one. Thank you, Howard, for your commitment and passion! Our excellent editors at Ballantine nurtured us with equal parts encouragement and rigor: Brendan Vaughan seized on our proposal with unfettered enthusiasm, envisioned the book it could become, and gave us clear headings to steer by; Sara Weiss skillfully guided the manuscript through multiple drafts, exhorting us at every turn to hone our storytelling to the max. Sara, you have earned our respect, admiration, and gratitude. To all our other angels at Ballantine, the Ross/Yoon Agency, and Hotchkiss Daily & Associates: our kudos and thanks.

Finally, we want to give a special shout-out to Stephen Mills, friend and fellow writer, who gave us the straight dope on what was working on the page, and what wasn't, at numerous stages along the way. Thank you, Stephen, for your sage and steadfast counsel.

I want to thank my friends, family, and colleagues who contributed their memories and reflections to this book. And to my son, Hakeem, my constant companion, Mario Kart partner, and intellectual sounding board through this past year of quarantine: my heartfelt thanks.

—Hakeem

Thanks to my Masala Art writers group, whose esprit de corps has leavened the book-writing grind. And first, last, and always, I'm eternally grateful to my North Star, Ericka Markman, who daily lights my way with love.

—Josh